Plants, Patients and the Historian

ENCOUNTERS cultural histories

Series editors:
Roger Cooter
Harriet Ritvo
Carolyn Steedman
Bertrand Taithe

Over the past few decades cultural history has become the discipline of encounters. Beyond the issues raised by the 'linguistic turn', the work of theorists such as Norbert Elias, Pierre Bourdieu, Michel Foucault or Jacques Derrida has contributed to the emergence of cultural history as a forum for bold and creative exchange. This series proposes to place enounters – human, intellectual and disciplinary – at the heart of historical thinking. *Encounters* will include short, innovative and theoretically informed books from all fields of history. The series will provide an arena for exploring new and reassembled historical subjects, stimulating perceptions and re-perceptions of the past, and methodological challenges and innovations; it will publish at history's cutting edge. The *Encounters* series will demonstrate that history is the hidden narrative of modernity.

Already published

Dust Carolyn Steedman

Plants, Patients and the Historian

(Re)membering in the Age of Genetic Engineering

Paolo Palladino

Rutgers University Press
New Brunswick, New Jersey

First published in the United States 2003
by Rutgers University Press, New Brunswick, New Jersey

First published in Great Britain 2002
by Manchester University Press
Oxford Road, Manchester M13 9NR, UK
http://www.manchesteruniversitypress.co.uk

Library of Congress Cataloging-in-Publication Data and
British Library Cataloguing-in-Publication Data are available upon request.

ISBN 0-8135-3237-X (cloth)
0-8135-3238-8 (paper)

Printed in Great Britain

Contents

List of illustrations

Acknowledgements

In many ways, *Plants, Patients and the Historian: (Re)membering in the Age of Genetic Engineering*, a history of genetics and the present, as well as a reflection on historiographical method and the philosophy of history, is a synthesis of work that has already appeared in the following articles and essays. The reader is referred to these articles and essays for more detail than will be found in *Plants, Patients and the Historian.*

'The political economy of applied science: Plant breeding research in Great Britain, 1910–1940', *Minerva*, 28 (1990): 446–68.

'Between craft and science: Plant breeding, mendelian genetics, and British universities, 1900–1920', *Technology and Culture*, 34 (1993): 300–23.

'Wizards and devotees: On the mendelian theory of inheritance and the professionalization of agricultural science in Great Britain and the United States, 1880–1930', *History of Science*, 32 (1994): 409–44.

'Science, technology, and the economy: Plant breeding in Great Britain, 1920–1970', *Economic History Review*, 48 (1995): 116–36.

'The empire, colonies and lesser developed countries as mirror: Critical reflections on science for economic development in the colonial and post-colonial periphery, 1930–1970', in Y. Chatelin and C. Bonneuil (eds), *Nature et Environnment* (Paris: ORSTOM, 1996), pp. 243–53.

'On writing the histor(ies) of modern medicine', *Rethinking History*, 3 (1999): 271–88.

'Icarus' flight: On the dialogue between the historian and the historical actor', *Rethinking History*, 4 (2000): 21–36.

'Speculations on cancer-free babies: Surgery and genetics at St Mark's Hospital, 1924–1995', in J.-P. Gaudillière and I. Löwy (eds), *Heredity and Infection: The History of Disease Transmission* (London: Routledge, 2001), pp. 285–310.

'Between knowledge and practice: On medical professionals, patients and the making of the genetics of cancer', *Social Studies of Science*, 32 (2002): 137–65.

All these articles and essays could not have been written without the financial support of the Economic and Social Research Council and the Wellcome Trust (grants R000232443; 032188/Z/90/Z; and 043145). I am very grateful to them. This said, I am especially thankful to Adam Bencard, Fred Botting, Roger Cooter, Bülent Diken, David Edgerton, Mary Fissell, Paul Fletcher, Jean-Paul Gaudillière, Philip Goodchild, Piers Hale, Jonathan Harwood, Rhodri Hayward, Matthew Hilton, Sally Horrocks, Jeff Hughes, Mark Jenner, Robert Kirk, Nikolai Krementsov, Kirsten McAllister, Adrian Mackenzie, Maureen McNeil, Ray Monk, Tiago Moreira, Malcom Nicolson, Simon Nightingale, John Pickstone, Marcia Pointon, Steven Pumfrey, Chris Renwick, Skuli Sigurdsson, Roger Smith, Thomas Söderquist, Steve Sturdy and Teresa Young. Simply put, *Plants, Patients and the Historian* would not have existed, even in my most private thoughts, if it had not been for our many arguments over the nature of both history and historiographical practice, and the many helpful suggestions for further reading received. Importantly, while John Beatty and Jim Griesemer, as well as Mick Dillon and Colin Thirtle, might find it hard to believe, what I have learned from them about philosophy and politics has helped me immensely in reading and arguing. Thank you. My final thanks go to Howard Caygill and Harriet Ritvo, whose reports on the penultimate draft of *Plants, Patients and the Historian* reminded me how difficult it is to keep the champions of the necessary and the contingent, the philosopher and the historian, simultaneously happy. You have spurred me to do better in seeking a perhaps impossible reconciliation.

Lastly, *Plants, Patients and the Historian: (Re)membering in the Age of Genetic Engineering* is dedicated to the memory of Lily Kay and the many conversations started, but never seen through to the end . . .

Introduction: life decoded?

> Tomorrow, the first rough draft of the human genetic code will be published – one of the epic achievements of contemporary science. We will know the gene sequences that determine our mental and physical behaviour. We will have the tools that in decades ahead will allow us to understand how much of what we do is predetermined and how much is of our own free will.

> The bourgeoisie cannot exist without constantly revolutionising the instruments of production, and thereby the relations of production, and with them the whole relations of society. Conservation of the old modes of production in unaltered form was, on the contrary, the first condition of existence for all earlier industrial classes. Constant revolutionising of production, uninterrupted disturbance of all social conditions, everlasting uncertainty, and agitation distinguish the bourgeois epoch from all earlier ones. All fixed, fast-frozen relations, with their train of ancient prejudices and opinions, are swept away, all new-formed ones become antiquated before they can ossify. All that is solid melts into air, all that is holy is profaned, and man is at last compelled to face with sober senses his real conditions of life, and his relations with his kind.

> It has not escaped our notice that the more we learn about the human genome, the more there is to explore.

Plants, Patients and the Historian: (Re)membering in the Age of Genetic Engineering is the result of nearly fifteen years of research into the role of genetics, the study of the genealogical patterns of plant and animal species, including *Homo sapiens*, in the reorganisation of British agriculture and medicine during the century just past. At the risk of misrepresenting Harriet Ritvo's more meticulous and measured work, I would say that this role is particularly prominent because the study of genealogical patterns has enjoyed a unique place in English, if not British, culture. In Britain, merit is still strongly linked to birth. Yet, as these words are being written, genetics also seems to be dramatically transforming our understanding of ourselves, including our relationship to the past and the future. *Plants, Patients and the Historian* is therefore also concerned with the relationship between the act of historical recollection and the coming 'age of genetic engineering', the millennial age to which the first epigraph speaks. Significantly, if this and all subsequent epigraphs will appear without attribution, at least in the first instance, it is because attribution begins to tell a story about authorship and context of production. I, the author of *Plants, Patients and the Historian*, may not be willing to concede such narratives, if only for a moment, to thus emphasise the tacit work done by the historian when confronted with nothing but a few words. Such work is not dissimilar to that done by the geneticist when confronted with a few plants or a few patients. In the meantime, I do hope that you will excuse my interpellating you, the reader, so directly, or my being so insistent on authorial presence, *pace* Roland Barthes' and Michel Foucault's pronouncements about the death of the author, as it is all part of the problems that *Plants, Patients and the Historian* seeks to confront.

Let me say a few words about the title, *Plants, Patients and the Historian: (Re)membering in the Age of Genetic Engineering*, however. The disaggregation of the word 'remembering' signals my interest in the role of the archive in historiography, the professional practice of connecting the present and past actions and events. Following Walter Benjamin's famous 'Theses on the philosophy of history', which will figure silently throughout *Plants, Patients and the Historian*, the archive can be understood as a site of remembrance. It is the site where the historian seeks to reintegrate those figures of the past that were at the centre of such actions and events from disconnected and dispersed fragments – the verbal, visual, auditory and tactile materials, which this historian would most probably call 'historical records'. Yet, if remembrance is then a form of 'restitution unto the dismembered' – I use the antiquated preposition 'unto' quite advisedly, to evoke a certain messianic tone – it is a reintegration that takes place in the present and is inescapably orientated toward the future. Remembrance is about death and redemption.[1] This, however, begs difficult questions about the exact relationship between the past, the present and the future. The phrase 'in the age of genetic engineering' is instead intended to echo Benjamin's other, equally famous, but quite different, essay on 'The work of art in the age of mechanical reproduction', which will also figure silently throughout *Plants, Patients and the Historian.* If Benjamin's 'Theses on the philosophy of history' speak of despair about the possibility of ever effecting any restitution, given the irresistible violence of the forces of 'progress', 'The work of art in the age of mechanical reproduction' instead raises the possibility that this violence might one day emancipate humanity from its enslavement under the rule of capital.

At the beginning of the twentieth century, a moment that marked both the culmination and crisis of the modern faith in the ability of humans to become the measure of all things, genetics promised to make the world anew. It seemed that the problems then besetting the theory of evolution by natural selection might be resolved by assuming that the visible and variable characteristics of the whole organism could be reduced to a mathematically exact recombination of hidden and invariant units of inheritance. In 1907, William Bateson, who first coined the word 'genetics' for this new approach to the patterning of resemblance and difference between generations, had no doubts that

> There is something that will come out of [genetics] that will equal, if not exceed, in direct consequence, anything that any other branch of science has ever discovered. . . . A precise knowledge of the laws of heredity will give man a power over his future that no other science has ever endowed him.[2]

For Bateson, genetics was the key to understanding the dynamics linking the past and the present, and thus determining the future. Yet, as Hannah Arendt first noted in her essay on 'the human condition', the conceptual categories and analytical framework of genetics are closely associated, if not intimately involved with the most fateful extension of the modern faith that humans could one day become the measure of all things. This was the genocidal madness of the Holocaust. The link is such that the celebrations of the recent decoding of the human genome have often been punctuated by recollections of the Holocaust. Thus, the first epigraph, drawn from an editorial in the *Observer* and aptly entitled 'The book of life: Gene science spells out our destiny', continues by also noting that:

> The moral and social implications are barely discussed. How much of our intelligence . . . is acquired and how much is endowed by our gene structure? In the early twentieth century the so-called science of eugenics informed Nazi arguments to justify the holocaust. Once again we are exposed to the risk that deadly value judgements will be linked to the structure of some gene sequences over others. If we can tell at birth that a child will, say, have criminal tendencies, what is the appropriate reaction?[3]

Any book that proposes to examine the historical role of genetics in the reorganisation of human practices as important as agriculture and medicine cannot therefore avoid addressing the ethical and political questions raised by the Holocaust. Yet, addressing such questions cannot amount simply to a discussion of how one should or should not recollect particular events in the history of genetics, because, however arguably and disputably, the entire intellectual fabric of Western thought was called into question by the Holocaust. My concern with the problems involved in remembering the making of the age of genetic engineering is then caught uneasily between Edith Wyschogrod's *An Ethics of Remembering* and Gillian Rose's *Mourning Becomes the Law.* Their views on the place of the Holocaust in contemporary philosophical discourse are fundamentally different. While the former views the Holocaust as introducing a radical rupture in our relationship to the past, the latter argues that the Holocaust represents just one moment, however dramatic, in the historical development of Western culture and its peculiar political institutions. Whatever the relative merits of Wyschogrod's and Rose's arguments may be, it remains none the less true that, for historians, a more thorough engagement with the questions raised by the Holocaust must begin with a return to the archive.

Admittedly, the late Irving Velody and others have re-evaluated the place of the archive in historiographical practice quite exhaustively, in a recent issue of *History of the Human Sciences.* Furthermore, one of the contributors to Velody's collection, Carolyn Steedman, has extended her reflections on the archive in *Dust,* the first volume to appear in 'Encounters'. The contribution of *Plants, Patients and the Historian,* the second volume to appear in this same series, is to note more modestly that the practices of genetics and modern historiography have much in common.

In his 'Theses on the philosophy of history', Benjamin suggested that what he called the 'cultural treasures' out of which history is written, and which the professional historian might dispassionately call 'archives', are nothing but the repositories of the victors' spoils in the struggle for political power. Similarly, Richard

Dawkins, the great populariser of genetics and evolutionary biology, never tires of reminding his audiences that the genome, the assembly of genes determining the characteristics of any one individual of a species, encapsulates the history of those genes that won out in the biological struggle for existence. Furthermore, the archive and the genome are both haunted by the traces of an incomplete erasure of the defeated, that is, by the contradictions in the archival repository and by the recessive alleles in an organism's genetic complement. Such haunting is the engine of both the endless need to revisit history and the equally endless extension and reformulation of genetic concepts to which the third epigraph speaks. Significantly, the report from which the lines of this third epigraph are extracted, a report by the International Human Genome Sequencing Consortium announcing the 'initial sequencing and analysis of the human genome', closes with the following lines from T. S. Eliot's 'Little Gidding':

> We shall not cease from exploration.
> And the end of all our exploring
> Will be to arrive where we started,
> And know the place for the first time.[4]

'Little Gidding', one of the poems constituting the famous *Four Quartets*, speaks of the modern subject's impossible longing for freedom from history. It speaks of the impossible longing to know a place so well as to be able to tell its past and future without the shadows of doubt that make life in the present dramatically uncertain.

Recalling the recently deceased Stephen Jay Gould, that other great populariser of evolutionary biology and persistent opponent of Dawkins' genetic determinism, and especially his brilliantly entertaining essay *The Panda's Thumb*, one might note further that these similarities between genetics and modern historiography are not a matter of analogy, but of homology. The practices of genetics and modern historiography are rooted in a common historical moment. As Michel Foucault argued in *The Order of Things*, this was the 'discovery' that the diverse phenomena that characterise the ever-changing world about us are reflections of an immanent logic of material necessity, the logic of 'life'. Genetics and modern historiography speak to this logic.

Significantly, although they speak in the distinctive registers that allow us to say that 'this is a historian' and 'this is instead a geneticist', genetics and modern historiography share one last, common aim. This is the realisation of that secular redemption that goes by the name of the 'Enlightenment'. Of course, defining this concept is notoriously difficult and treacherous. As will become evident in the course of *Plants, Patients and the Historian*, my definition is informed, at least partly, by Foucault's essay 'What is enlightenment?' and all of its ambiguous, if not contradictory, implications for our understanding of Foucault's place in the history of Illuministic thought. My initial contention then is that exploring the similarities between the geneticists' and historians' practices can serve very effectively to illustrate how the archive, like the plants or patients mobilised by the agricultural or medical geneticist, is not simply a repository of the past. It is also the principle of formation of the past, the present and the future. I would also contend that it is the principle of formation of both the historical actor that was and the historian or geneticist that will be. As Paul De Man has put it more pithily than I could ever do,

> The power of memory does not reside in its capacity to resurrect a situation or a feeling that actually existed, but is a constitutive act of the mind bound to its own present and oriented toward the future of its own elaboration.[5]

Thus, the argument of *Plants, Patients and the Historian* might be that the past, present and future, as well as the relationship between the 'self' and the 'Other', are made in the archive, botanical, medical and historical. Yet, Karl Marx and Friedrich Engels' famous and much quoted phrase 'all that is solid melts into air, all that is holy is profaned', should also call for pause and reflection.[6] Ultimately, *Plants, Patients and the Historian: (Re)membering in the Age of Genetic Engineering* is an attempt to articulate such a reflection.

Notes

1 See also Derrida, *The Gift of Death*, pp. 1–34, esp. p. 13.
2 Bateson, 'Toast of the Board of Agriculture', p. 76.
3 Anonymous, 'The book of life', p. 28.
4 Eliot, as quoted in Lander *et al.*, 'Initial sequencing and analysis of the human genome', p. 914.
5 De Man, *Blindness and Insight*, p. 92.
6 Marx and Engels, 'Manifesto of the Communist Party', p. 338.

1

Genes, archives and history

> I trust in youth that has led me aright when it now compels me to protest at the historical education of modern man, and when I demand that man should above all learn to live, and should employ history only in the service of the life he has learned to live.

> A Klee painting named 'Angelus Novus' shows an angel as though he is about to move away from something he is fixedly contemplating. His eyes are staring, his mouth is open, his wings are spread. This is how one pictures the angel of history. His face is turned toward the past. Where we perceive a chain of events, he sees one single catastrophe which keeps piling wreckage upon wreckage and hurls it in front of his feet. The angel would like to stay, awaken the dead, and make whole what has been smashed. But a storm is blowing from Paradise; it has got caught in his wings with such violence that the angel can no longer close them. This storm irresistibly propels him into the future to which his back is turned, while the pile of debris before him grows skyward. The storm is what we call progress.

Agriculture, medicine and genetics

Long before the intellectual crisis encapsulated by contemporary discussions of the Holocaust, and the many questions these discussions have raised about the nature of the Enlightenment, Friedrich Nietzsche observed that modern historiography itself was complicit with the destructive impulses of the Enlightenment. The first epigraph, one of the many mesmerising phrases in Nietzsche's essay 'On the uses and disadvantages of history for life', summarises Nietzsche's call on historians to engage with the present, for the sake of 'life'.[1] Today's public debates over the genetic modification of the food we eat and the introduction of genetic tests to prevent our offspring from suffering unnecessarily present those historians who might wish to meet Nietzsche's challenge to write a history 'in the service of life' with a unique opportunity. What would seem to be at stake in these debates is the future of 'life' itself.

As Paul Rabinow has noted, in his provocative essay on 'Artificiality and enlightenment', the meaning of pivotal terms such as 'nature' and the 'subject' is becoming deeply problematic as genes are isolated, appropriated, and moved across species, including that most important of animal species that is *Homo sapiens.* Even the idea of 'history' is called into question. The point is neatly captured by Kate Douglas, a writer for the *New Scientist*, while reporting on the renewed arguments between champions of genetic determinism such as Richard Dawkins, and those who, following Stephen Jay Gould, revel instead in the unpredictability and indeterminacy of evolutionary history. She writes:

> Evolution is history. Not dead, terminated, finito, you understand, but rather the unfolding story of life on Earth. And as in any history, chance plays a role. Just as it is impossible to know if the First World War would still have started if Archduke Francis Ferdinand's driver had not made a wrong turn, so it is impossible to say what life would have been like if, for example, the age of dinosaurs had not ended in catastrophe. Or is it? Some evolutionary biologists think there is more to evolution than mere history.[2]

The gist of these last evolutionary biologists' iconoclastic argument is that, thanks to developments in the field of 'artificial life', it is now empirically demonstrable that, if evolutionary history is replayed, the outcome is largely the same. Evolutionary history is thus the unfolding of an immanent logic, the logic of the gene, which then means that it is within geneticists' reach to alter and determine its future course. Arguably, the achievement of such millennial power is the logical consummation of the Enlightenment.

If we are to credit Giorgio Agamben, it would appear that classical culture drew a categorical distinction between *zoē*, the material form of existence shared by humans and other living creatures, and *bios*, the ethical and political life unique to the citizens of the *polis*. Sometime between the seventeenth and eighteenth century, however, this pivotal distinction began to be erased. Michel Foucault, one of the most acute critics of the Enlightenment, illustrated such erasure by reproducing in *Discipline and Punish* the frontispiece to Nicolas Andry's *Orthopaedia* (see figure 1). In this frontispiece, which evokes a world where human agency is no longer, as it begs questions about who exactly bound the tree, all difference between the straightening of misshapen trees and the physical and moral education of children, with whom the orthopaedist was once concerned, is erased. In other words, these formerly disparate practices are reduced to common principles of a literally disembodied and all-pervasive instrumental reason. Foucault is, of course, famous for his sustained discussion of how 'biology', the science of 'life' that has systematically implemented such reasoning to erase all difference between *zoē* and *bios*, eventually acquired its current, central role in the organisation of Western society. Oddly, and perhaps significantly, the enterprise should then be labelled 'zoology' rather than 'biology'. While Foucault, the historian, did not inquire into the development of this so misnamed enterprise beyond its initial, formative period, other historians have endorsed his insight, even when they are not immediately concerned with his critique of the Enlightenment. For example, Garland Allen has discussed how, during the early twentieth century, the disparate practices of plant breeding and eugenics, the selective reproduction of

Figure 1 From Nicolas Andry, *Orthopaedia: Or, the Art of Preventing and Correcting Deformities in Children* [1741] (London, 1743)

the human species, were reduced to common genetic principles, which left little room for the classical distinction between *zoē* and *bios*. This reductive assimilation is clearly conveyed by a poster issued by the Eugenics Society, which called on its contemporary viewers to only sow 'healthy seed' (see figure 2).

Admittedly, by the time of this poster's production, sometime around 1935, plants and humans were not quite the same thing, but it was only a matter of a very short time before they became the same. As Zygmunt Bauman has noted in *Modernity and the Holocaust*, the assimilating logic of biology was fully realised within the next twenty years, in that consummation of the 'gardening' principle that was the Holocaust. Significantly, while the frontispiece to Andry's *Orthopaedia* begs questions about the location of agency, in the Eugenics Society's poster, agency is clearly invested in an individuated, but featureless and generic, male human subject, who is to take responsibility for where and what 'oats' he 'sows'. Equally significantly, even though almost everyone would agree with the injunction to 'never again' repeat the fatal elisions that led to the Holocaust, the denial of difference between *zoē* and *bios* implied by the poster's imagery continues to this day, more tacitly, but none the less resolutely and invasively.

For example, genetic engineering is today challenging the social institutions that once maintained the culturally unstable boundaries between humanity and the natural resources it uses to nourish the body. In Britain, the Agricultural and Food Research Council was disbanded just about fifteen years ago, and the then newly established Biotechnology and Biological Sciences Research Council subsumed most of the former institution's activities. At the time, similar changes in the Medical Research Council were also mooted, since most of the research it supported was deemed to be equally biological. Thus far, however, the medical profession has successfully resisted the merger of the Medical Research

Figure 2 Reprinted with the permission of the Galton Society and the Contemporary Medical Records Archive. Courtesy of the Wellcome Library, London

Council and Biotechnology and Biological Sciences Research Council. Strikingly, such resistance is not based on some fundamental differentiation between humans and the plants or animals that once were the concern of the Agricultural and Food Research Council. It is instead based on the pragmatic argument that the Agricultural and Food Research Council and the Medical Research Council had evolved along such different lines that any merger is now impracticable. It is thus obvious that the anchors that secure the separation of the world of things, or 'nature', and ourselves, the quintessential embodied 'subject', are being severely tested, if not yet fully weighed. In fact, as Howard Caygill has noted, this situation is resulting in the convergence between the most disparate positions one might imagine. The British Medical Association's report *Our Genetic Future* and the Papal encyclical *Evangelium Vitae* would appear to concur in arguing that more biological research is the answer to the problems raised by the materialisation of being realised by the decoding of the human genome. Art is not immune to this momentous convergence, as performance artists such as Stelarc announce,

> The end of the Darwinian concept of evolution through organic change. From now on, with nano-technology, mankind can absorb technology. Thus the body must no longer be considered as the seat of the spirit or as the instrument of human relations but must be considered as a structure. Not as an object of desire, but as an object to be redesigned.

Almost as if wishing to recall Andry's reductive assimilation of plants and humans, Stelarc then imagines a future human, who is endowed with photosynthetic skin. He/it writes,

> With such a skin we would no longer have need of a mouth to chew, of a throat to swallow, of a stomach to digest, of lungs to breathe. We would be able to leave the human and replace useless organs with technologies. Ha-ha-ha-ha-ha-ha-ha-ha-ha-ha-ha-ha! . . .[3]

It might then seem that there is little more to be said about the relationship between agricultural and medical applications of genetics, and about the more general project of the Enlightenment. All that remains for the historian to do is to fill in the historical details, perhaps by noting how Stelarc's vision echoes an earlier futurist aesthetic of the machine, and then argue that what today seems absolutely unprecedented is in fact no such thing.

Public responses to contemporary genetic innovations in the fields of agriculture and medicine seem to suggest, however, that the apparently well-established, converging relationship between *zoē* and *bios* may perhaps be less stable than one might think. The British public clearly fears genetically modified foods. Writing their own history, by invoking more nefarious images from the formative period of the Enlightenment, such as those conjured in Mary Shelley's *Frankenstein*, they, or, to be more precise, that *vox populi* that is the *Daily Mail*, convey their fear of genetically modified foods with the expression 'Frankenstein foods'.[4] At the same time, this same public views the supposedly parallel genetic modification of human beings more ambivalently. This situation is captured by two responses from the virtual audience attending a discussion, which was held during a recent instalment of the BBC television programme 'Question Time', on the subject of 'designer babies', the expression that often encapsulates popular response to the genetic modification of human beings.[5] The first response sought to rebut widespread criticism of genetic technology by highlighting the benefits of research into the genetic bases of human disease:

> Gene therapy is a powerful technology that has the potential to help many people with illnesses and diseases. People with weak heart muscles, cystic fibrosis, degenerative discs, spinal injuries, and a whole host of other diseases could benefit from this new scientific medical breakthrough.

The second respondent, however, could not understand how the public could mistrust the agricultural applications of genetic technology, and, at the same time, be so uncritical of its medical applications: 'It appears GM humans are okay, but GM crops are not, I cannot here [sic] the protest.'[6]

One might dismiss the public's differentiation between the genetic modification of food and the body, to which this last respondent referred, as the myopic product of the increasing historical amnesia noted by Frederic Jameson and other critics of contemporary culture. It overlooks, for example, how Shelley's monster was in fact the result of modifying the human body, not food crops or livestock. One might also point out that the differentiation forgets that more practically significant and notorious attempt to control the human genetic constitution that resulted in the Holocaust. The balance of fears, in other words, should at least be inverted. Alternatively, however, one could admit that there are in fact serious distinctions between the agricultural and medical sciences, which critiques of the Enlightenment tend to overlook as they unreservedly condemn its reductionism.

Differentiating agriculture and medicine

The grounds for differentiating agriculture and medicine are manifold and complex. Although it might offend the environmentalist sensibilities of the members of Earth First, as well as Bruno Latour's philosophical sensibilities, one might argue that plants can be bent at will. Admittedly, the wheat plant has not yet been modified so that it can do better than just produce its own supply of toxins to fend off insect pests or suffer no ill effects from the use of Roundup Biactive™ as a weedkiller. We still do not have a wheat plant that can approximate a *perpetuum mobile* by producing its own supply of nitrogen. This is not because the wheat plant has any power to resist the will of genetic engineers, but because these engineers have not yet fully understood its genetic constitution. It is just a matter of time before they will have overcome this problem. At the same time, however, many members of the British public worry that, despite the reassurances of public officials and scientific advisory committees, genes inserted into agricultural crops will break loose and become a new and far more devastating source of environmental pollution. As environmental activists and growing numbers of organic farmers have warned for the past few years, these genes will drift onto organic crops and thus deprive consumers of their freedom to choose how they should nourish their bodies. Moreover, these consumers find it too hard to believe that such potentially dangerous modification has anything to do with providing the less fortunate than themselves, the hungry and the poor all over the world, with cheap food, as is often claimed by the same public officials and scientific advisory committees. It seems more likely that the only ones to benefit will be the powerful, and until recently largely invisible, agri-chemical companies and food manufacturers who have done most to advance the development of genetically modified food crops. Of course, these concerns echo the bitter controversies over the genetic innovations that marked the now largely forgotten 'Green Revolution', which Vandana Shiva has described as nothing less than a neo-colonialist conspiracy to deprive the poorest people of the world of their 'natural' rights. The present genetic modification of food crops brings the expropriation closer to home, and many members of the public are so alarmed that they turn to the expression 'Frankenstein foods' as a way of representing the perceived threat to their own natural rights.

The situation with regard to the genetic modification of the human body seems quite different. Patients certainly experience the world of medicine as though they were plants, that is, as incapacitated, silent objects. They submit none the less so that they might become once again subjects in full possession of all their faculties. Patients also accept their objectification because, even if it fails to return them to subjectivity, it contributes to the benefit of humanity. To put the matter more pointedly, when people donate their blood and sometimes even their organs, or their dying children's organs, they are not helping to build a new monster, but believe instead that they are helping others to return to subjectivity. In Britain, where blood has not yet become a commodity, such donations are viewed as acts of humanity in a world that all too often seems utterly inhuman. More problematically, patients may also believe that they cannot be bent at will because they, unlike plants, are endowed with the power to resist the objectifying logic of the Enlightenment and that, moreover, the agents of that logic would be changed as well. As Dietrich von Engelhardt has argued, the critic of the Enlightenment frequently overlooks how medical professionals often protest as much as their patients against the modern organisation of medicine and its denial of subjective experience. As he points out, it is the medical practitioner who often articulates the critiques of the inhuman modern civilisation in the great novels of modernist literature, from Marcel Proust's *Remembrance of Things Past* to Virginia Woolf's *Mrs Dalloway*. Today, the critique is articulated more prosaically, by railing against 'evidence-based medicine'. Thus, when the medical practitioner experiments with, or appropriates a dying child's organs without his or her parents' permission, it is not medicine itself that is identified as the problem. The problem is instead the all too corruptible human, inattentive to the fundamental dilemma of modern medical discourse, namely, deciding when objectification is legitimate and when it is not. That is to say, the development of new therapeutic practices always demands the treatment of an ailing or dying patient as an object, sometimes fruitfully and sometimes not. This, however, must be weighed impossibly against the hallowed Hippocratic injunction to do no harm, which problem now provides the legitimation for the rapidly growing field of medical ethics, the professional discourse normalising and policing approaches to this fundamental dilemma of modern medical discourse.

There is, of course, a historical literature that calls this understanding of modern medicine into question. As Roger Cooter has noted in his critique of medical ethics, the latter is a thinly disguised effort to protect medical professionals' exclusive power over the ailing or dying patient. Much can be said in favour of such scepticism. The defensive reaction of the General Medical Council to reforms of its organisation, proposed largely in response to the recent public uproar over medical specialists' surgical experiments and use of dying children's organs without the parents' permission, would certainly support it. Perhaps, however, the greater limitation of medical ethics is that it presumes to resolve rationally the troubling question at the heart of medical discourse. Arguably, we have learned from the Holocaust that ethics should instead be a discourse that transcends rational calculation, for one's rationality was proved on this occasion to be another's unconscionable irrationality. More importantly for the purposes of *Plants, Patients and the Historian*, the imputed political motivation of medical ethics has not had a counterpart in agriculture, at least until quite recently. Even now, though, this counterpart is only concerned with animals, perhaps because we are occasionally prepared to extend to them some form of consciousness, if not of subjectivity. Significantly, this recent extension of ethical discourse is again replete with imponderable ambiguities as we negotiate our way between subjective relationships with our pets and calls for the humane treatment of the animals we eat. I certainly find it very difficult to reconcile rationally my love for my two dogs, Rachel and Moses, and my fondness for a filet mignon *au bleu*, carved out of an anonymous cow. I cannot but envy my vegetarian partner's greater coherence.

In sum, something strange happens when one enters a relationship with another subject, which produces a world of difference between the discourses of agriculture and of medicine in their relationship to the logic of the Enlightenment.

Historians, historical actors and the archive

If historians are to provide some critical insight into the complexities of the contemporary development of genetic engineering, they must attempt to synthesise the histories of agricultural and medical science. Significantly, in his essay on 'the question concerning technology', which was written in the shadow of the Holocaust, Martin Heidegger observed that science and technology were thoroughly undermining the autonomy of the human subject. However, Heidegger's now very topical observation was part of a more general argument about the limitations of instrumental reason, which also called attention to the complicity of modern practices of representation with this dehumanising historical process. A critical synthesis of the histories of agricultural and medical science should then be attentive to its possible complicity with the same process. Yet, the current relationship between historical studies of the place of the agricultural and medical sciences in the reorganisation of British agriculture and medicine during the nineteenth and twentieth centuries would appear to mimic that between the Agricultural and Food Research Council and the Medical Research Council. These studies have developed independently of one another, but share common modes of representation.

If the Holocaust has not been particularly influential in the thinking of historians, philosophers and sociologists of science of the now passing generation, the same cannot be said about the prospect of nuclear warfare and global annihilation. For many of these historians, philosophers and sociologists, science could no longer be viewed as the key to the secular redemption promised by the Enlightenment. Thus, studies of the agricultural and medical sciences ceased to be celebrations of the great men, and occasionally the great women, who changed the world through their genius and astute deployment of scientific rationality in understanding the structure and functions of plants and the human body. Thomas Kuhn's *Structure of Scientific Revolutions* instead provided the philosophical rationale for the separation of historians' accounts of the developments of such understanding from the accounts offered by scientists past and present, accounts which appealed to indisputable, objective facts of nature and the equally indisputable logic of instrumental rationality. The authority of nature was replaced by the authority of the archive, the repository of those records excluded from scientific publications and other recollections of scientific achievement. The archive became the preferred source for a more sceptical evaluation of scientific progress. Furthermore, according to the sociological approach that took the place of logical empiricism in the aftermath of *Structure of Scientific Revolutions*, the historical development of the natural sciences was to be explained causally by the historically specific social, political and cultural contexts in which it took place. The contrast between the two approaches divided by the crisis of confidence in science is perhaps conveyed most resonantly by the radically different accounts of the origins of genetics provided by William Provine, on the one hand, and by Donald Mackenzie and Barry Barnes, on the other hand. If the resolution of the problems that beset the relationship at the beginning of the twentieth century between genetics and the theory of evolution by natural selection, was, for Provine, a matter of individual intellectual genius, for Mackenzie and Barnes, it was a matter of professional politics and eugenic considerations.

Significantly, Steven Woolgar eventually criticised the historicist approach exemplified by Mackenzie and Barnes' study, an approach often labelled as 'social constructionism', for its failure to be 'reflexive'. In his innovative essay on 'interests and explanation', Woolgar argued that, if the truths of the natural sciences were partial, so were the truths of the social sciences. This criticism then impelled a radical transformation of method among a number of sociologists and anthropologists interested in science, technology, and medicine. In *Laboratory Life*, Woolgar and Bruno Latour joined forces to extend the former's argument and demand a 'symmetric' treatment of the analytical categories deployed by natural and social scientists.

Most famously, between the two editions of *Laboratory Life*, the subtitle changed from *The Social Construction of Scientific Facts* to the *Construction of Scientific Facts*. This was because, in the course of advancing the proposed, more symmetric approach, the ontological status of the material objects of scientific inquiry became deeply problematic. The issue was raised most pointedly in Latour's *The Pasteurization of France*, a study of the emergence of the 'germ theory' of disease, in which Latour fully articulated his theory of the 'actor-network'. In this study, the moving matter visualised by the microscope was endowed with a role fully comparable to that of the human beings and social institutions involved in the emergence of the 'germ theory'. More recently, in *Le Pouvoir des Malades*, Vololona Rabeharisoa and Michel Callon have pursued the approach further, to cast a wholly different light on the role of patients in the production of something as arcane as the map of the human genome. The network of participation in this momentous feat is so distributed across the boundaries between the material and the social worlds as to effectively reduce the notion of human agency to an artefact of the network. In Sharon Traweek's still more thoroughly anthropological approach, attentive to the ambiguous position of the 'participant–observer', the very possibility of the social scientist representing scientific developments as a removed observer, however much this social scientist might be sceptical about the categories of classical sociology, is called into question even further. The act of representation becomes inseparable from the object of representation, and this is no more obvious than in Donna Haraway's widely celebrated *Modest Witness @ Second Millennium: FemaleMan© Meets OncoMouse™*.

This said, most historians of science, technology and medicine have paid little attention to the implications of these arguments for their own practices of representation. For example, in his recent and promising historical study of food research in the Rowett Research Institute, David Smith has noted that agricultural researchers and medical researchers espoused the same managerial ideology that characterised many of the connections between science, technology and society since the First World War. Yet, the actual organisation of agricultural research, at least as it took shape in the Rowett Research Institute, was very different from that of medical research.[7] Smith fails to note whether this difference is in any way related to differences between the objects of agricultural and medical research. Perhaps this is because historians' relationship with their own subject of inquiry is so much more removed than is the case with sociologists and anthropologists, who can witness directly all the tacit elisions, transgressions and epistemological challenges that characterise the practices of scientific research. All that historians have before them are silent archival fragments. Historians then continue to engage with the archive in a scientific manner that mirrors the very same problematic approach taken by the scientists they are studying. They do so even when they engage with the critiques of science, natural and social, and turn to 'deconstruction', as in Tim Lenoir's collection of essays on *Inscribing Science*. It is striking how this promising turn, with all its emphasis on bringing into existence objects such as the nucleic acids and proteins in Hans-Jörg Rheinberger's outstanding essay 'Experimental systems, graphematic spaces', is limited to the scientists' worlds. The turn does not extend to the historians themselves and the world they bring into existence. Lenoir, in his introductory essay to *Inscribing Science*, rightly notes the difficulties presented by 'deconstruction' for conventional forms of historical narrative, but, with the sole exception of Hans Gumbrecht's concluding essay, the problem is not explored any further. Gumbrecht's essay offers scant consolation, however, for he ends simply by throwing up his hands, writing that,

> It is high time to put aside the illusion that perceptions will ever be completely captured by concepts and reflections. Instead of relying upon that illusion ever being fulfilled and coming to an end, one simply has to be able to give up.[8]

Ironically, although informed by Jacques Derrida's writings on deconstruction, Gumbrecht fails to note both the productive nature of this 'illusion', including the production of *Inscribing Science*, and what might be the sources of this so productive 'illusion'.

An altogether different response to the problem of historiographical representation is characterised by Thomas Söderquist's criticism of Adrian Desmond and James Moore's popularly and critically acclaimed biography of Charles Darwin. Söderquist has argued that, by reducing Darwin and his insights into the order of the natural world to passive reflections of the prevailing social organisation and politics of Victorian society, Desmond and Moore effectively denied the disruptive, if not revolutionary, impact of Darwin's insights. More importantly, however, Söderquist has also highlighted the ethical problems that are raised by such objectification of historical actors and has demanded an alternative historiography that will return subjectivity to the historical actor, to the moral benefit of the observing historian and his or her readers. The point is undoubtedly important, for the price Gumbrecht asks of historiography is too high, but Söderquist never explains the nature of the links that would make the necessary empathy between the historian and the historical actor at all possible. He instead tacitly assumes a shared sense of embodiment and subjectivity, into which he falls easily since his particular biographical subject is still very much alive. As Söderquist puts it,

> With an increasing number of interviews with Jerne [Niels Jerne is the subject of Söderquist's forthcoming biography], after 50-100 hours of discussion, the notions of seemingly endless variability of texts, the 'death of the author', and the non-referentiality of the proper name became increasingly absurd.[9]

Söderquist clearly overlooks the fundamental difference between coming to know Jerne and knowing Jerne. The former is a contextualised practice, at the end of which stands the common phrase 'you can never fully know someone', begging questions about the origins and ontological status of the object of the practice. More importantly, however, all that most historians encounter are in fact not embodied humans, with whom we might share a tacit sense of self-identity and continuity. They encounter instead archival fragments, that is, disparate and disconnected verbal, visual, audible or tactile materials, which taken individually and in themselves are as mute as the geneticist's plants or animals.

Coming to terms with these methodological difficulties requires a fundamentally different understanding of the relationship between the historian and the archive.

The archive and the coming into being of the historian

Once found, it is relatively easy to read archival fragments and from them then write an objectifying history of agricultural genetics.

These archival fragments are usually silent about the relationship between geneticists and their objects of inquiry, plants or animals. At best, they suggest a relationship in which these plants or animals respond to the questions posed by the geneticist in an unambiguous manner determined by the unilaterally imposed experimental conditions. One would not expect anything different given the most common assumptions about the nature of plants and animals. Histories of agricultural genetics then rehearse the presumed relationship between plants or animals and the geneticists of the past who articulated our present understanding of genetics, by denying the geneticists any agency in the moment of historiographical construction. That is, geneticists are only allowed to answer the historian's questions in the most unambiguous manner, and have no recourse against his or her will to explain events in the past differently than they themselves might have done. The historical development of genetics is not explained by experiments with plants and the application of scientific rationality to understand the results of such experiments, but it is explained instead by the geneticists' social circumstances. Thus conceived, the geneticists are not political 'subjects', but 'objects' comparable to the geneticists' plants or animals, and our relationship to the geneticists is equally monological. This is especially evident in those exceptional circumstances when the geneticists are still alive and actively participate in the historiographical construction. As Betty Smocovitis has noted, questions are then raised about the merits of the constructions, as the historian comes to be seen as partisan, if not as someone as unreliable as a journalist. The monologue then seems to be a crucially important feature of the proper relationship between the historian and the historical actor, or, to more accurately reflect historiographical practice, between the historian and the historical actor whom we imagine to have left in the archive traces of him or herself.

By contrast, archival fragments concerning medicine and genetics are often seen as replete with contradictions about the relationship between the imagined agent, for example, the clinicians interested in the inheritance of a human disease, and their object of inquiry, humans affected by this disease. Because these last figures are imagined by the historian to be human beings rather than plants or animals, there is a shared experience of embodiment and subjectivity between these humans and the historian, which commands alertness to such contradictions and then impels a more politically, if not ethically engaged historiography. Such special attentiveness is perhaps necessary because these patients' imagined agency is vitally important to sustain one's own sense of agency within contemporary medical discourse. That is, imagining patients who protest against the medical profession is important for shaping one's present and future, as one confronts the promises of genetic engineering with trepidation. Thus, historians of medicine such as the late Roy Porter have shown great interest in patients' narratives and in their opposition to the medical profession's claims to authority. The problem, however, is that these historians have often approached patients' narratives and opposition by deploying the very same objectifying methodologies deployed in the study of agriculture. The critical voices of the imagined patients are reduced to the rhetorical repertoire of a historicist narrative that resolves the contradictions between medical professionals' desire to objectify and patients' opposition. In striving for such closure, however, these historians effectively undermine the political, if not ethical, thrust of their work. To put this claim more pointedly, one might imagine the patient who was once silenced by the objectifying logic of medicine calling on the historian to speak for him or her. Those politically and ethically engaged historians who respond to this call then silence this patient anew by focusing on those social structures that hid him or her from historical consciousness. Revealing and explaining the silence thus silences and denies agency all over again. Significantly, Barbara Duden's *Geschichte unter der Haut*, mistranslated quite significantly as *The Woman Beneath the Skin*, is perhaps the notable exception to this account of the place of the patient in the history of medicine. However impossibly, this sensitive anthropological study of a medical practitioner in early modern Germany seeks to avoid any interpretation.

All these problems are very familiar to feminist historians such as Carolyn Steedman, for whom methodology and her subject of inquiry, the historical place of 'woman', are inextricably connected. Not surprisingly these historians have found themselves at the forefront of the critique of modern historiography. As Joan Scott's collection of essays on *Feminism and History* amply illustrates, however, these historians are also deeply divided on how far to pursue the critique, for it seems politically disempowering as it threatens to dissolve the very category 'woman' that motivates their endeavours. Perhaps one needs to re-examine not just one's understanding of the relationship between the historian and the archive, but one's understanding of the politics and ethics of responsibility to the 'other' as well.

The historian is clearly distinguished from the archive because he or she does not simply represent archival fragments, but rather driven by erotic desire reconstructs the past in all its integrity, lovingly suturing such fragments together to bring the historical actor back to life, and perhaps return agency to them. Of course, the interpretative thread for such suturing is not of the past, but originates in the present, in the historian's understanding of what it means to be a human being rather than a plant. The suturing with such thread may sometimes be so overwhelming, however, that the imagined historical actor ceases to be a subject. Like the plant or animal in the hands of the agricultural geneticist, this historical actor becomes instead a puppet moved by strings visible to the historian alone. The historian can love too much and then become a puppeteer. In other words, our interpretative practices in the archive can be complicit with the logic of the Enlightenment that would submit everything to instrumental rationality, however much we may in fact aspire to the contrary. What is required is some kind of symmetry between the historian and the historical actor, where agency, if any longer meaningful, falls somewhere between the two.

Such symmetry is possible if one understands the archive as a site of *aporia* or as a *punctum.* That is, the archive is the site of an irresolvable and disruptive question, such as that presented by the puzzling frontispiece to *Nothing New Under the Sun*, a collection of essays encountered during this study of remembrance in the age of genetic engineering. As will become evident, this frontispiece marks uneasily the boundaries to human desire to become the measure of all things (see figure 3).

For Roland Barthes, the *punctum* is 'the element which rises from the scene, shoots out of it like an arrow, and pierces me . . . this wound, this sting, speck, cut, little hole'.[10] This wounding, a founding

Figure 3 From J. P. Lockhart Mummery, *Nothing New Under the Sun* (London: Andrew Melrose, 1947).

violence, then initiates a chain of hermeneutic and transforming practices whose aim is to return the subject to its original integrity, however impossibly, as resolution is always referred to another, yet undiscovered, archive. Thus one can view the archival encounter as effectively calling the historian into being as the *punctum* and *aporia* are uneasily closed by historical narrative. The archive then becomes the site of engagement between the simultaneously and mutually constituted 'self' and 'other'. Moreover, since the past, present and future also come into being in this moment, the historian's political orientation toward the future does not in fact pre-exist the encounter with the archive, but is as much a textual effect as the historical actor or agent conjured in the archive. Understanding historiography in these performative, rather than representational or mimetic terms, opens the possibility of a more meaningful, if not 'lively', engagement of the historian with the present. The archive no longer presents the historian with lessons from the past, or, more importantly, with stages of an unfolding historical logic. The archive instead initiates a relationship between the historian and the historical actor that is as testing as that which one, like any other member of the public, has to confront as one engages with the world of agriculture and medicine today, not fully knowing what tomorrow holds in store. This is, perhaps, what Friedrich Nietzsche meant when called on historians to write histories 'for life'.

Historiography as process and performance

It is also clear, however, that the Nietzschean turn, however conceptually compelling and emancipating, also entails loss. It entails the surrender of the messianic aspirations that have lain at the heart of the historiographical imagination since Giambattista Vico, and impelled Walter Benjamin's melancholic portrayal of the historian's impossible desire to 'stay, awaken the dead, and make whole what has been smashed'.[11] The 'aura' of the authentic, unique and irreproducible moment is lost, and historical narrative becomes no different from that advanced by the 'evolutionary biologists' cited by Kate Douglas in her article in the *New Scientist.* We might say that history becomes simply the passage of time, marked by births, lives and deaths, and that its reversibility is now within our grasp, thanks to the promise of genetic engineering. As Richard Dawkins has recently announced, 'in fifty years we could resurrect the past'.[12] Yet, it is worth recalling that the report by the International Human Genome Sequencing Consortium announcing the 'initial sequencing and analysis of the human genome' also closed by noting that 'the more we learn about the human genome, the more there is to explore'. Perhaps, to understand this contradictory vision of the end of history, we need to attend much more carefully to the meaning of 'life' as such.

We might begin by recognising that there may be something excessive about being human that is not easily reducible to the statue conjured by François Jacob, one of the chief architects of that impressive construction that is genetics today, in his aptly entitled autobiography, *The Statue Within.* I wish to suggest that the approach best suited to the challenge is best captured by Michel De Certeau's recollection of Jean de Labadie, in the concluding chapter of *The Mystic Fable*:

> A man of the South . . . Labadie went north . . . From Guyenne, where he was born and became a Jesuit, he went to Paris, Amiens, Montauban, Orange . . . then thought perhaps he would go to London, no, it was Geneva, then the Netherlands, Utrecht, Middleburg, Amsterdam, then farther, to Altona in Denmark, where he died ... The inner journey was transformed into a geographical one. Labadie's story is that of indefinite space created by the impossibility of place. The stages of the journey are marked by the 'religions' he passes through, one by one: Jesuit, Jansenist, Calvinist, Pietist,

> Chiliast or Millenarian, and finally 'Labadist' – a mortal stage. He passes on. He cannot stop.[13]

For Labadie, an unsociable deconstructionist *avant la lettre*, truth was affirmed in the ever more manifest impossibility of comprehending it.

In sum, one might then say that the truth sought by historiography and genetics, the truth of 'life', is affirmed in the very act of dialogical engagement and refusal of any truth that seems within the realm of positive determination. Truth may lie in the process of speaking to something that, in the process, becomes evermore manifestly unknowable. Let me then begin to substantiate this allusive and elusive argument by revisiting a narrative about the emergence of 'Frankenstein foods', which, for a brief, all too brief moment, seemed to me definitive.

Notes

1 Nietzsche, 'On the uses and disadvantages of history for life', p. 116.
2 Douglas, 'Replaying life', p. 29.
3 Stelarc, as quoted in Caygill, 'Liturgies of fear', p. 102.
4 For an introduction to the figure of Frankenstein and its place in the history of genetics and genetic engineering, see Turney, *Frankenstein's Footsteps.*
5 For an introduction to the historical origins of the phrase 'designer babies', see Squier, *Babies in Bottles.*
6 http://www.bbc.co.uk/tv/: 'Question Time', 5 October 2000.
7 See Smith, 'The use of "team work" in the practical management of research in the inter-war period', esp. pp. 278-80.
8 Gumbrecht, 'Perception versus experience', p. 364.
9 Söderquist, 'Existential projects and existential choice in science', p. 59.
10 Barthes, *Camera Lucida*, p. 33.
11 Benjamin, 'Theses on the philosophy of history', p. 249.
12 Dawkins, 'The word made flesh', p. 11.
13 De Certeau, *The Mystic Fable*, p. 271.

~ 2

Plants, genetics and the modern state

> There is something that will come out of [genetics] that will equal, if not exceed, in direct consequence, anything that any other branch of science has ever discovered. . . . A precise knowledge of the laws of heredity will give man a power over his future that no other science has ever endowed him.

Remembering beginnings

Nearly fifteen years ago, I joined the Centre for the History of Science, Technology and Medicine, in the University of Manchester, as a research associate on a project on the history of plant breeding in Britain. The project had been designed by Jonathan Harwood, a historian of genetics, as well as, to me at least, a leading social constructionist, and Colin Thirtle, an agricultural economist interested in technological innovation.

At the time of my arrival, Jonathan was working toward the completion of *Styles of Scientific Thought*. In this book, he argued that, in inter-war Germany, the diversity of zoologists' and botanists' views on the scope and significance of genetics could be explained in terms of these botanists' and zoologists' distinctive social backgrounds. In *The Role of Demand and Supply*, Colin approximated Jonathan's argument that the evolution of scientific knowledge is shaped by social institutions, by arguing that technological innovation is not a factor that is external to economic calculation, but is instead reducible to the dynamics of economic supply and demand. Given their intellectual, as well as institutional proximity, Jonathan and Colin then thought that they should join forces to obtain funding from the Economic and Social Research Council, under its recent initiative to foster social and economic studies of the process of technological innovation. They succeeded.

I took up Jonathan and Colin's project as a historian of ecology and agricultural science who knew virtually nothing about either British history or social constructionism. I was interested in Stephen Jay Gould and Richard Lewontin's celebrations of historical contingency and their politically charged critiques of genetic reductionism, however.[1] Over the subsequent five years, I immersed myself in institutional and social histories of Britain during the twentieth century, and interviewed many academic researchers, breeders, and businessmen. Of course, I also covered myself in grime in musty attics, trying to piece together the history of the then recently privatised Plant Breeding Institute, 'a unique national asset' whose sale was sometimes compared to selling the 'crown jewels'.[2] Prompted by the criticism of an anonymous referee, I must also add that these 'archives' were particularly grimy. This was because one of the features of institutions such as the Plant Breeding Institute was that their place in British society was so assuredly entrenched, *pace* what did in fact happen to them, that they had no care for the past. Moreover, their mission was forward-looking. Their sole aim was to provide for the most mundane, but necessary, of human needs, by putting on the table food that was always cheaper than yesterday's dinner. Thus, the archival records of these institutions' past often were little more than randomly assorted piles of papers, which no one was particularly sure whether they should be thrown out or not. These archival records then usually ended up in damp attics. What began to emerge out of the consequent, filthy rummaging is the following history of not just the Plant Breeding Institute, but also of its chief institutional rivals, the Scottish Plant Breeding Station and the Welsh Plant Breeding Station. Needless to say, it is difficult to remember accurately the thoroughly objective spirit in which this history was written, but let me try to remember and re-tell.

As I discovered in the course of those five years of work, during the second half of the nineteenth century, the elimination of the Corn Laws, tariffs on agricultural imports into Britain, opened new markets for British industrial exports in an increasingly competitive global market. As a result, however, the economy of rural Britain declined steeply.[3] Landowners raised rents, making the life of tenant farmers increasingly difficult, and both shed agricultural labourers to secure their dwindling profits. Farms were abandoned, and unemployed labourers migrated into urban centres. Conservatives, aiming to secure the continuing vitality of their largely rural political constituencies, then began to press for the reintroduction of tariffs. Liberals, however, opposed such tariffs, fearing the cost of economic retaliation abroad and the political effects of more expensive food at home. Ironically, British farmers were vastly more efficient than their competitors overseas – Britain was the home of not just the 'industrial revolution', but of the 'agricultural revolution' as well. The compromise between the Conservatives and Liberals was to improve rural infrastructures, so as to make these farmers even more efficient. In 1909, the Liberal government introduced the Road Improvement and Development Fund Act, which provided unprecedented public funding for the extension of railways and the building of new roads throughout rural England, Wales, Scotland and Ireland. Significantly, Henry Dale, a senior civil servant who was intimately involved in the subsequent evolution of the Development Commission, once recalled that the Chancellor of the Exchequer, David Lloyd George, introduced the bill by arguing that,

> A state can and ought to take a longer view and a wider view of its investments than can individuals. The resettlement of deserted and impoverished parts of its own territories may not bring to its coffers a direct return which would reimburse it fully for its expenditure; but the indirect enrichment of its resources more than compensates it for any apparent and immediate loss. The individual can rarely afford to wait, a State can; the individual must judge of the success of his enterprise by the testimony given for it by his bank book; a State keeps many ledgers, not all in ink, and when we wish to judge of the advantage derived by a country from a costly experiment, we must examine all these books before we venture to pronounce judgement.[4]

The proposed social 'experiment', and its implicit surbordination of the individual to the needs of the state, attracted much opposition, but it was eventually accepted. Yet, to ensure that the Liberals would not use the funds provided by this legislation to further undermine Conservative dominance in rural Britain, the House of Lords sought to invest a small number of Development Commissioners, accountable to the Privy Council alone, with the responsibility of administering the new funds. The consequences of this delegation of authority were momentous.

The parliamentary debates on the Road Improvement and Development Fund Act had focused largely on the manner of expropriating landowners to build railways and roads. The newly appointed eight commissioners instead focused their energies on the promotion of agricultural research, a previously overlooked, minor aspect of the Road Improvement and Development Fund Act. This was perhaps unsurprising since one of the commissioners was Sidney Webb, the influential advocate of professional expertise as the solution to social and political problems confronting the British Empire. It is unclear whether Webb was particularly responsible, but the Commission was especially attracted to the new science of genetics. Heredity had figured prominently in many discussions of eugenic answers to the problems of the Empire, and Lloyd George certainly referred to these problems and answers while defending the Road Improvement and Development Fund Act against Conservative opposition. William Bateson, who first coined the word 'genetics', argued much more ambiguously that the most promising application of this science was in improving the productivity of British farmers. As he put it, in a toast to the Board of Agriculture during the Third International Conference on Genetics, which was held in London two years before the Parliamentary wrangles over the Road Improvement and Development Fund Act,

> There is something that will come out of [genetics] that will equal, if not exceed, in direct consequence, anything that any other branch of science has ever discovered. . . . A precise knowledge of the laws of heredity will give man a power over his future that no other science has ever endowed him.[5]

Between 1912 and 1923, the Development Commission then proceeded to establish the Plant Breeding Institute in Cambridge, the Welsh Plant Breeding Station in Aberystwyth, and the Scottish Plant Breeding Station just outside Edinburgh. The aim of these three centres was to promote the development of genetics to replace traditional practices of plant breeding, such as the search for 'sports', the occasional, unexplainable appearance of outstanding plants, and 'selection', which aimed to propagate the characteristics of the sports throughout the crop. Genetics would thus help to erase the historically produced differences between one plant and another, and to establish thereby more uniform and economically valuable crops. During the next sixty years, the programme upon which the Development Commission thus embarked proved so far-reaching that the three plant-breeding centres became important targets in the redefinition of the role of the state in British society advanced by the Conservative Prime Minister Margaret Thatcher.

In 1986, the Conservative Deputy Chief Whip of the House of Lords, the Earl of Swinton, defended the proposed privatisation of the Plant Breeding Institute by reversing Lloyd George's understanding of the responsibilities of the state. He responded to the government's critics by stressing that,

> It is Government policy to encourage industry funding of R&D, to improve technology transfer from publicly funded research to industry for the benefit of the UK recovery and to consider privatisation where this might foster sound management, commercial efficiency and good value for money for the taxpayers. It is therefore entirely appropriate that the position of the PBI [Plant Breeding Institute] should be reviewed in this way.[6]

Significantly, the funding by the Economic and Social Research Council, which enabled Jonathan and Colin's project, was just one aspect of this effort to obtain 'value for money'. Anyway, the Secretary for Science and Education, Kenneth Baker, pursued the Earl of Swinton's argument further by arguing that,

> The Government are satisfied that transferring [the Plant Breeding Institute] to the private sector is the best way of ensuring that [its] experience and expertise is applied and developed to the fullest extent and to the maximum benefit of the United Kingdom agriculture and the United Kingdom economy.[7]

On the eve of the explosion of corporate interest in genetic engineering, the Plant Breeding Institute was sold to Unilever for £66 million, which was over three times the initial valuation by the accounting firm Lazard Brothers. Ten years later, Unilever sold what had become Plant Breeding International to Monsanto, for the even more staggering amount of £320 million. As the *Financial Times* put it, the amount paid by Monsanto was quite 'surprising' given that the activities of Plant Breeding International had not changed substantially during the years since its privatisation.[8] Monsanto was perhaps speculating on the future of genetic engineering, just before the furore over what the *Daily Mail* labelled, with its usual panache, 'Frankenstein foods'. The argument of this chapter is that the contours of this future were set in place almost eighty years earlier, with the enactment of the Road Improvement and Development Fund Act and the establishment of the Plant Breeding Institute, the Welsh Plant Breeding Station and the Scottish Plant Breeding Station.

The business of breeding

During the hearings over the Road Improvement and Development Fund Act, an agreement that the beneficiaries of any capital grants from the Treasury should provide matching funds sought to further allay the fears that the Act might become a vehicle for the expansion of a growing bureaucratic state. It was never clear, however, who exactly was to be the ultimate beneficiary of the grants.[9] Strikingly, the most prominent actors outside the government to become involved in the establishment of the Plant Breeding Institute, the Welsh Plant Breeding Station and the Scottish Plant Breeding Station were not farmers, but seed firms, millers and brewers.

During the late nineteenth and early twentieth centuries, brewers were among the most politically influential businessmen, and figures such as Lord Iveagh, the owner of Guinness Brewers, were deeply involved in the reform of British society. As the scale of their businesses expanded nationally and internationally, the control of quality became increasingly important. Large and uniform shipments of barley were particularly attractive, but British farmers simply could not match their Canadian and American competitors since they provided much smaller lots of barley, whose malting qualities varied enormously from one lot to another. Given the size of their businesses, the largest brewers could negotiate with Canadian and

American grain merchants for favourable prices, but smaller ones could not. The latter, speaking mainly through the Institute of Brewing, were then very interested in the promotion of greater crop uniformity among the British growers from whom they preferred to buy their barley. A similar problem faced wheat farmers, who could not provide the largest British millers, such as Huntley and Palmers, Hovis or Spillers, with large and uniform lots of wheat that were of any use for the making of bread flour. Organisations such as the Home-Grown Wheat Association were then actively involved in the organisation of the plant-breeding centres envisioned by the Development Commission.

At the same time, however, the proposed plant-breeding centres presented a potential threat to seed firms. On the one hand, the most significant hurdle confronting innovative and internationally renowned seed firms such as Gartons, Carter's Seeds and Sutton's Seeds was the absence of any protection of breeders' rights over their products. Although the British Seed Corn Association, an organisation of breeders and seed firms established in 1903, realised that one solution might have been to enact legal protection of proprietary rights, for many of its members this was not an acceptable solution. As long as anyone succeeded in producing a new and attractive variety, individual members of the seed trade wished to preserve open access to such a variety. Moreover, as the secretary of the renamed National Association of Corn and Agricultural Merchants pointed out in 1921, during the negotiations over the establishment of the Scottish Plant Breeding Station, any attempt to introduce seed registration amounted to nothing less than state interference with the market. An alternative solution was to transfer the cost of plant breeding onto the state. The screening and comparison of existing varieties to establish their economic value and assess the heritability of different economically important characteristics, crucial and expensive preliminary activities of plant breeding, could be immensely useful. Furthermore, any publication of the data accumulated in field trials could allow seed firms to identify 'synonymous' varieties and thus those unscrupulous members of their trade who were marketing varieties developed by others. Publicly funded plant-breeding research could thus act in lieu of legal protection without the unwelcome constraints associated with such legislation. While smaller unscrupulous seed firms could be excluded by informal means, the larger ones could continue their practice of selling synonymous varieties without fear of incurring any penalty. On the other hand, these larger seed firms also were very wary of publicly funded breeders producing new varieties that might be placed on the market and displace their own. There was much discussion of modelling the plant-breeding centres after the Sveriges Utsädesförening. Under this Swedish scheme, varieties produced by a partly publicly and partly privately funded research centre at Svalöf were turned over to a joint stock seed company, which was responsible for marketing them. Profits were used to meet the costs of research and breeding, and any residue was shared among the shareholders. However acceptable in Sweden, the organisation of a seed company closely associated with a publicly supported research station was unanimously rejected by the British seed trade.

Given the determination to avoid any semblance of outright state intervention, by imposing the condition that beneficiaries of public grants should match them, and the conflicting priorities of some of these tacit beneficiaries, the three plant-breeding centres were then configured very differently, depending on the contingencies of their specific location. In the early twentieth century, Britain still was a largely decentralised nation. Genetics, however, played a subtle, but none the less important, role in changing all this.

The Plant Breeding Institute

In the late 1880s, the newly established county councils of rural England sought to give agricultural education a firmer institutional foundation, and they hoped that the University of Cambridge might train the teachers needed to help them to do so. Over the next decade, however, the university opposed the development, arguing that education in technical and commercial subjects was incompatible with academic education. Yet, the Colleges constituting the University of Cambridge were themselves some of the largest landowners in Britain, and they thus had a clear interest in putting some remedy to the agricultural depression. Thus, in 1899, the University of Cambridge accepted the offer of the Worshipful Company of Drapers to endow a chair in agriculture. Significantly, this endowment was conditional on the organisation of a course leading to a degree in agriculture, rather than the less academically demanding diploma, as was the case in the departments of agriculture in other universities and agricultural colleges. In 1912, following the inauguration of the first plant-breeding research centre supported by the Development Commission, the Plant Breeding Institute, as a department of the recently established School of Agriculture, agricultural studies gained still greater academic respectability. As Thomas Barlow Wood, the second Drapers' Professor of Agriculture, pointed out, the teachers in the School would now be doing more than providing the heirs of the landed gentry with a scientific education. They would have sufficient funds to enjoy the 'time-honoured tradition in Cambridge that every teacher in the science schools should engage in research in his own subject'.[10] This opportunity for research, moreover, came with no strings attached since the condition that beneficiaries of funding from the Development Commission should match the funds was waived, presumably because the

University of Cambridge already paid the salaries of researchers in the Plant Breeding Institute.

Soon after the First World War, seed and milling firms exploited the continuing tensions over the place of agriculture within the University of Cambridge to establish the National Institute of Agricultural Botany. Seed firms, in particular, were concerned that the Plant Breeding Institute might become a competitor for control over the market, and argued therefore that plant-breeding research should be separated from the development of new crop varieties. Attention to commercial activities, the seed firms claimed, would be detrimental to the scientific programme of the Plant Breeding Institute. This claim was strongly supported by both academics in the University of Cambridge and the Development Commission. It was eventually agreed that the National Institute of Agricultural Botany should oversee the commercial development of 'elite' seed stock, that is, the genetically pure lines that lie at the basis of modern hybrid varieties, and then market any new 'finished' varieties produced by the Plant Breeding Institute. Even this, however, seemed to concede too much to the Plant Breeding Institute. The board of the National Institute of Agricultural Botany insisted that varieties developed by the Plant Breeding Institute, unlike those developed by private breeders, should not be placed on the market until it could be shown that they were clearly and distinctly superior to existing ones. The policy restricted researchers in the Plant Breeding Institute to screening existing varieties and, according to Frank Engledow, effectively 'discouraged' them from developing new ones.[11] Arguably, such screening was much more important to the patrons of the National Institute of Agricultural Botany. As the numerous varieties of wheat available on the market were screened to evaluate the heritability of economically important features, such as their suitability for the making of bread flour, the data was used to call public attention to the merits of the different varieties and to identify 'synonymous' ones. In fact, the publication of the findings of the Synonym Committee of the National Institute of Agricultural Botany was very effective in eliminating those firms that were simply multiplying and selling others' varieties under their own proprietary names. The National Institute of Agricultural Botany thus acted to some extent to ensure protection of the traders' proprietary rights over the varieties they placed on the market.

This institutional arrangement freed research workers at the Plant Breeding Institute from having to pursue lines of work that were primarily or exclusively of immediate and practical significance. Researchers in the Plant Breeding Institute and the School of Agriculture, such as Engledow and Redcliffe Salaman, the latter of whom is remembered today mainly as the author of the historiographically innovative *The History and Social Influence of the Potato*, were regarded as more interested in statistics and genetics. In fact, they were closely involved in organising the publication of the *Journal of Agricultural Science*, whose editorial policy was to decline papers dealing with 'farming as opposed to agricultural science'.[12] Farming questions were so unimportant that very little attention was paid to the details involved in producing commercially viable varieties, and the results thereof were sometimes quite embarrassing. For example, 'Steadfast' and 'Holdfast' wheats, which were presumed on theoretical grounds to be well suited to the needs of the milling industry, failed to gain any wide acceptance because Engledow had overlooked facts that 'everyone knew', everyone being the Home-Grown Wheat Association.[13]

The work of the Plant Breeding Institute, however, was not altogether disconnected from the world of commercial plant breeders. For example, its first director, Sir Rowland Biffen, stressed continuously the importance of genetics to the development of plant breeding by citing the widespread adoption of his 'Yeoman' wheat. Biffen's successor, Herbert Hunter, was not convinced, however, that knowledge of genetics was so important. Reflecting on his earlier experience working for Guinness Brewers, Hunter argued that commercially important features, such as yield and milling quality, were not inherited in a fashion according with the mendelian laws of segregation, the conceptual foundations of genetics. Contrary to Biffen and other advocates of genetics, it seemed to him that it was virtually impossible to breed according to any 'preconceived plan'.[14] William Bateson was equally sceptical about the relationship between genetics and Biffen's achievements as a breeder, privately casting doubt on Biffen's actual understanding of genetics. Assuming the accuracy of Hunter and Bateson's assessments of Biffen's work, it then seems that a practically minded breeder, rather than a geneticist, as the Development Commission would have liked, directed the Plant Breeding Institute. This was perhaps the price to pay to reach a compromise between the seed, milling and brewing firms, the University of Cambridge, and the Development Commission.

The Scottish Plant Breeding Station

The establishment of the Scottish Plant Breeding Station just outside Edinburgh took an altogether different form, which, in many ways, reflected Scottish farmers' very different response to the agricultural depression that had plagued Britain since the 1870s. While English farmers had turned arable land over to grass and sought to minimise capital investment, Scottish farmers extended the length of crop rotations and intensified the cultivation on the thus reduced productive acreage. One effect of this different approach was that they were far more dependent on good seed than English farmers were. They found, however, that a large portion of the seed on the market was unreliable. There was therefore considerably greater interest in Scotland in the services of an institution that could test the viability and purity of the seed. The Development

Commission, however, resisted calls for its support because they were not interested in such an institution. The impasse was broken during the First World War.

Wartime governmental price supports encouraged farmers everywhere to produce more, and this led them to pay greater attention to the seed they purchased. Demands for government support for seed testing became more insistent, and acquired greater legitimacy in the wake of an official report, written by the Development Commissioner Thomas Middleton, to the effect that Britain, as a whole, had become over-reliant on imported agricultural products, including agricultural seed. These demands soon resulted in the creation by the wartime Food Production Department of an Official Seed Testing Station in Cambridge. After the war, the disgruntled seed firms made sure that the National Institute of Agricultural Botany subsumed this institution. In the meantime, the advocates of a Scottish institution dedicated to the improvement of agricultural seed pointed out that Scotland could not but depend on seed imported from Sweden because testing in Cambridge was not helpful for Scottish farmers. Climatic and soil conditions in East Anglia and Scotland were too different for any field tests in the former region to be of any relevance in the latter. This protestation was ignored until a substantial number of Scottish seed firms and the owners of larger land-holdings – fifty acres or more – constituted the Scottish Society for Plant Breeding Research, with the intention of establishing a plant-breeding centre along the lines established by the Sveriges Utsädesförening. Presumably, because this private initiative threatened the work of the Development Commission, the Development Commission and the Department of Agriculture for Scotland then agreed to lend their support. The original, formal objections that the work done by the Plant Breeding Institute met national needs was nullified by restricting the range of activities of the Plant Breeding Institute and the proposed Scottish Plant

Breeding Station to locally important crops. Research in the Scottish station was to emphasise potato and oat breeding, while the Plant Breeding Institute would attend to wheat and barley breeding, effectively ignoring that wheat and barley crops were just as important in Scotland as they were in England. While the agricultural geography of the nation was subtly transformed, however, the politically more sensitive testing and certification of seeds would remain unified and cover all crops.

Significantly, the commercial origins of the Scottish Plant Breeding Station were reflected by the expectation that its maintenance would depend on profits from the sale of new varieties. This expectation was not unanimously shared. Many of the larger members of the seed trade were unhappy about the prospect of a well-capitalised competitor. The Scottish Society for Plant Breeding Research settled therefore on a compromise. The Scottish Plant Breeding Station would offer 'elite' seed stock only to subscribing members of the Society, and on the basis of payment of a flat fee rather than royalties. The purchasers would then take responsibility for multiplication and sale. The Scottish Plant Breeding Station was not associated with a university, which might pay the researchers it employed, however. It then came to depend exclusively on the interest from the initial endowment grant from the Development Commission and the Department of Agriculture for Scotland, the annual subscriptions paid by the members of the Scottish Society for Plant Breeding Research, and payments for 'elite' seed stock. Consequently, the Scottish Plant Breeding Station often was very hard pressed for funds, especially as the fortunes of the agricultural economy reached their nadir in the inter-war years and demand for 'elite' seed stock dried up.

The commercial orientation of the Scottish Plant Breeding Station also shaped its research programme. For example, the Development Commission had identified the improvement of plant varieties with genetics, and hence it expected that the directors of the plant-breeding centres it supported would be familiar with this science. Yet, the first director of the Scottish Plant Breeding Station, Montagu Drummond, had little such knowledge, but he was nevertheless preferred to geneticists from the School of Agriculture in Cambridge. He was instead an experienced taxonomist and plant pathologist. The latter expertise accorded with the interest of the Scottish Society for Plant Breeding Research in maintaining the high quality of Scottish seed potatoes, which were very attractive throughout Britain because Scottish climate was judged to be less conducive to the development of fungal and viral diseases. Drummond's expertise in taxonomy was instead useful in sorting through the bewildering number of varieties of potatoes found on the market. Furthermore, Drummond himself insisted that systematic breeding was impossible without a clear idea of varieties available for breeding, and without an equally clear idea about the heritability of their most desirable features. He also suggested that, under these circumstances, a certain amount of 'haphazard crossing' was inevitable, if not actually desirable.[15] It would then seem that, at least while he was director of the Scottish Plant Breeding Station, Drummond was committed first and foremost to the improvement of crops, not to genetics. The pattern was repeated when Drummond resigned in 1925 to take up the much more academically respectable chair of botany at the University of Glasgow. Drummond's former assistant, William Robb, was appointed as Drummond's replacement. Robb was an experienced potato breeder who lacked any advanced training in genetics or botany, his highest qualification being a National Diploma of Agriculture from the West of Scotland College of Agriculture. Applications by graduates of the School of Agriculture in the University of Cambridge, trained by Sir Rowland Biffen, were ignored once again. Robb's appointment made it abundantly clear that the Scottish

Society for Plant Breeding Research was not interested in having botanists, let alone geneticists, as directors of the Scottish Plant Breeding Station, but in dedicated plant breeders. Ultimately, however, the Scottish Society for Plant Breeding Research was far more concerned with sorting out potato varieties and establishing the resistance of existing seed stock to disease, than with producing new varieties. In sum, the organisation of the Scottish Plant Breeding Station was much closer in spirit to the National Institute of Agricultural Botany than to the Plant Breeding Institute.

The Welsh Plant Breeding Station

While plant-breeding research in Cambridge and Edinburgh was captive to academic and commercial interests, it began to take its contemporary form, as the fulcrum of a comprehensive reorganisation of rural Britain, in the most paradoxical of places. It began in the barren highlands of Wales, with the organisation and development of the Welsh Plant Breeding Station.

The University College at Aberystwyth was established to serve the local community. Its Agricultural Department, for example, received funds from the rural county councils to provide lectures on subjects such as butter-making and cheese-making. Although many academics in the University College thought that such instruction was beneath the dignity of a university, the College's dependence on the farmers' goodwill meant that there was little opportunity but to accede to the demand. In the early years of the twentieth century, these academics looked therefore to an increasingly interventionist state to support their position against that of the councils and the farmers represented by these councils, but to little avail. In 1912, however, the Board of Agriculture agreed to appoint an agricultural botanical adviser for the region surrounding Aberystwyth. This adviser would be attached to the Agricultural Department of the

University College. George Stapledon, a graduate of the School of Agriculture in the University of Cambridge, was chosen for the post and was immediately assigned the production of a report on the state of the seed trade in the region. Appearing on the eve of the First World War, the report concluded that most of the seed was unreliable because it often was adulterated and ill-suited to the needs of local farmers, mostly small-scale dairy farmers. Stapledon and academics in the University College increased their calls, asking for the establishment of a plant-breeding centre in Aberystwyth. The Development Commission, however, rejected the request, again because it believed that the Plant Breeding Institute was sufficient to serve the needs of the entire nation. Stapledon's wartime work as the director of the Official Seed Testing Station, however, lent greater legitimacy to his increasing criticism that the Plant Breeding Institute could not in fact produce varieties that were appropriate to the conditions in which Welsh farmers worked. Immediately after the war, Stapledon turned to Laurence Weaver, a renowned modernist architect, former director of the Food Production Department, and, at the time, president of the Imperial War Memorials Commission, for assistance. Together, they convinced the steel and shipping magnate Lord Milford to donate to the University College funding sufficient to establish a Welsh plant-breeding centre. This donation was again matched by a grant from the Development Commission.

Stapledon was the first director of the Welsh Plant Breeding Station, holding the post for the twenty-three years up to 1942. Throughout these years he dedicated his attention to the improvement of grasslands. In the most strictly technical terms, his approach to plant breeding was ecological rather than genetic. Stapledon's aim was to improve the productivity of pastures over the whole course of the season. He accomplished this by developing mixtures of grass species that would result in a continuous and high output of fodder. Improving the capacity of selected varieties to withstand intensive grazing and competition between species was, for Stapledon, a far more important problem than the development of the genetically pure lines established by Sir Rowland Biffen in the Plant Breeding Institute. Notwithstanding his own ecological approach, however, Stapledon did not discourage members of his staff from pursuing strictly genetic research, as was preferred by the Development Commission. In fact his close associate, T. J. Jenkin, undertook almost exclusively genetic investigations to improve the productivity of grasslands. There is no reason to believe therefore that the approach preferred by Stapledon was solely determined by the nature of the biological problem he confronted. It was equally likely that the approach was shaped by Stapledon's interest in the economic welfare of Welsh farmers, which he viewed, however, from the perspective of a human ecologist, observing the inextricable interaction of grassland, grazing livestock and human actions. It is perhaps not surprising then that he was the author of one of the first textbooks on 'human ecology' published in Britain. More importantly still, he refused to accept the idea that turning arable land into grassland and declining to invest in the improvement of highlands, in other words, not using fully the productive potential of the land, was a rational policy. What was needed was planning that took a long-term, ecological perspective, on the development of agriculture. The corporatist state, he argued, had to take the lead in demonstrating the effectiveness of this alternative perspective. Thus, Stapledon's 'Cahn Hill' experiment in ploughing the highlands was supported by grants from both the Development Commission and ICI, the leading manufacturer of artificial fertilisers anywhere in the British Empire. If necessary, Stapledon argued, the state should even return to David Lloyd George's pet project, the nationalisation of the land.

Stapledon's controversial ideas shaped the organisation and administration of the Welsh Plant Breeding Station. He was very critical of the firms selling seed to Welsh farmers, and thought that the state should discipline them by participating in the market and by setting the standard of service expected of them. He proposed therefore that the Welsh Plant Breeding Station should market its own varieties rather than turning them over to the seed firms through the National Institute of Agricultural Botany. In 1927, with the aid of two of the leading proponents of the corporatist state, Leo Amery and Sir Walter Elliot, Stapledon obtained financial support from the Empire Marketing Board to establish such a scheme. With the collaboration the Welsh Agricultural Organisation Society, he then proceeded to establish 'seed-growers' associations' that would take care of the multiplying and then marketing the crop varieties developed by research workers in the Welsh Plant Breeding Station. The latter retained the right to regulate and inspect the sowing of seed, the multiplication and mixing of certified crops, the post-harvest treatment of the seed, and lastly, to fix its price on the market.

Seed firms, of course, did not welcome the scheme. Thus, even though backed by the Empire Marketing Board, the Ministry of Agriculture and Fisheries resisted the implementation of the scheme. It argued that it was improper for academics such as Stapledon to participate so actively in matters of commerce. By 1938, however, state regulation of the economy was no longer as controversial. Stapledon's scheme was extolled by Political and Economic Planning, one of leading associations of corporatist, technocratic planners, as an example of progressive agriculture that would rescue Britain economically. In this more congenial political environment, Stapledon was finally allowed to form Aberystwyth Seeds Ltd. The company would sell 'elite' seed stock to nationally renowned seed firms, such as Dunns, but on the payment of royalties rather than the flat fees paid to the National Institute of Agricultural Botany and the Scottish Society for Plant Breeding Research.

Significantly, while the Welsh Plant Breeding Station produced a number of new varieties of herbage, few of them were widely adopted by the Welsh farmers who Stapledon sought to lead. These varieties were especially designed for intensive grassland farming. Given the economic circumstances during the inter-war years, however, Welsh farmers, like farmers throughout the rest of Britain, were reluctant to invest the capital required by Stapledon's intensive farming. In fact, in 1935, one such farmer confronted Stapledon, asking him bitterly what to do with the surplus milk produced under the latter's schemes. More milk was the last thing desired by the Milk Marketing Board, which at the time was seeking to reduce the national output of dairy products. It was not until the Second World War that Aberystwyth Seeds Ltd. began to generate a considerable income for the Welsh Plant Breeding Station, as Agricultural Executive Committees began to boost agricultural production by enforcing practices of intensive farming. Stapledon's revolutionary achievement was officially recognised by bestowing him with a knighthood.

Genetics and the nationalisation of agriculture

During the 1920s, the Development Commission acted as an advisory body to the Ministry of Agriculture and Fisheries and to the Treasury, with whom rested the final authority over the disbursement of the funds authorised under the Road Improvement and Development Fund Act.[16] As the early history of the Plant Breeding Institute, the Scottish Plant Breeding Station, and the Welsh Plant Breeding Station suggests, however, there was considerable tension over the exact relationship between the Development Commission, the Ministry of Agriculture and Fisheries, and the agricultural industries. These tensions were resolved by a series of local arrangements, based on local balances of institutional power. By the end of the 1920s, however, academic researchers such as those in the Plant Breeding Institute began to call for a more formal separation between the research institutes funded by the Development Commission and the Ministry of Agriculture and Fisheries. They argued that the latter's economic interests were incompatible with the proper conduct of scientific research. Plans were then drawn up to establish an Agricultural Research Council, comparable in its organisation to the Medical Research Council, which was not accountable to the Ministry of Health, but to the Privy Council alone. More specifically, they argued that the proposed Agricultural Research Council should only be concerned with 'the fundamental sciences that serve the development of agriculture', and should be 'mainly composed of men of high scientific standing in one or other of the basic sciences on which the ... development of agriculture depends'.[17]

According to the advocates of the Agricultural Research Council, plant-breeding research had to become research into general phenomena and processes of plant physiology, ecology and genetics. Not surprisingly, officials in the Ministry of Agriculture and Fisheries felt that justifying the policies of the proposed council before their nominal political constituency, farmers, would be difficult. The outcome of the ensuing struggle was a compromise. While the Agricultural Research Council, like the Medical Research Council before it, was to be placed under the administrative control of the Privy Council rather than Ministry of Agriculture and Fisheries, the latter would still retain executive control over the funding of agricultural research. The fundamentally different understanding of agricultural research in The Agricultural Research Council and the Ministry of Agriculture and Fisheries was nowhere more evident than during the discussions about the appointment of Sir Rowland Biffen's successor at the Plant Breeding Institute. In 1936, Sir Daniel Hall, the very influential, second director of the John Innes Horticultural Institution, the leading British centre for genetic research, was called by the Agricultural Research Council to comment upon Biffen's nomination of a former breeder for the Guinness Brewing company, Herbert Hunter. Hall opposed it on the grounds that Hunter was 'a plant breeder and not a geneticist, and has little interest in the important fundamental basis of a plant-breeding programme'.[18] The advice was over-ruled in this particular case, since the association with Guinness Brewers was too important to ignore, but Hall's philosophy came to dominate national policies for agricultural research as the Agricultural Research Council gained greater independence from Ministry of Agriculture and Fisheries. As Tim De Jager has pointed out, the strong academic orientation of the Agricultural Research Council was reflected in the biography of the Council's first secretary, Sir William Dampier. Dampier had been a physicist in the University of Cambridge, who had argued that the financial rewards of scientific research were unpredictable, and the Agricultural Research Council could not therefore undertake to deliver practical results to order. Such views were strongly criticised by renowned commercial plant breeders such as Edwin Sloper Beaven, who repeatedly challenged the dependence of plant breeding on genetics. Dampier, however, could count on the rhetorical skills of the first secretary of the Medical Research Council, Sir Walter Morley Fletcher, who, during a broadcast for the BBC, succeeded in transforming Beaven into a champion of publicly funded agricultural research, and especially of genetic research. A number of Dampier's successors tended to display their predecessor's scientific credentials. They were academic scientists, and, if anything, they enjoyed stronger connections to medicine than to agriculture. In their view, however, this was not an obstacle.

Echoing the frontispiece to Nicolas Andry's *Orthopaedia*, they shared a vision that drew no fundamental distinctions between the improvement of agriculture and health care. These were all fundamentally biological endeavours, and research scientists like themselves were uniquely prepared to plan their improvement. As Fletcher pointed out in 1930, farmers, like the despised medical men on Harley Street, knew as much about the scientific principles underlying their activities as did 'lawyers', the experts in unprincipled bargaining.[19]

Arguably, this attitude should be understood as instantiating what Andrew Abbott would call the process of academic 'expertization'. It is striking, however, how the academic scientists in the Agricultural Research Council were not particularly impressed by Biffen's production of 'Yeoman', which greatly satisfied the needs of the Home-Grown Wheat Association and was repeatedly celebrated by these academic scientists as the exemplary case of the economic benefits of unfettered scientific investigation. They were instead far more impressed by James Gregor's studies of ecological genetics at the Scottish Plant Breeding Station, despite the scantiness of resources, equipment, and funds provided by an avaricious Scottish Society for Plant Breeding Research. Gregor's approach, which entailed collaboration across academic disciplines, was much more consonant with the expansive understanding of agricultural science advanced by the Agricultural Research Council. Like Sir George Stapledon, and foreshadowing the ideas of that other former expert in economic development that is Bruno Latour, many of these scientists viewed human involvement in agriculture as amenable to scientific analysis, in the strictest sense of the word. To them, the transformation of nature and society were inextricably related, and only they knew how. Thus, for example, they believed that any input from agricultural economists, who might have been viewed as having a better understanding of society, was undesirable since their advice on costing and the like was inevitably tainted by their subservience to farmers' local, limited vision of the possibilities of agriculture. The real issue was the transformation of agriculture into an integral, if not central, component of a more advanced and planned national economy.

During the Second World War, the Ministry of Agriculture and Fisheries re-established the interventionist policies to ensure maximum output that it had first put in place during the previous global conflict. At the same time, however, it also began to plan for the expansion and intensification of post-war agricultural production, in an attempt to establish self-sufficiency once and for all time. This gave the leaders of the Agricultural Research Council an opportunity to implement their technocratic vision of agriculture. Reflecting the momentous political changes that had been taking shape in other areas of the national economy, they argued that improving efficiency and productivity required a centrally planned approach to the application of the latest scientific advances. The Welsh Plant Breeding Station, by this time the largest of the three plant-breeding centres and one of the largest of any research institutions supported by the Development Commission, thanks mostly to its considerable income from Aberystwyth Seeds Ltd, came to be seen as the model. To do this, however, the leaders of the Agricultural Research Council had to wrest from the Ministry of Agriculture and Fisheries all control over agricultural research, leaving to the Ministry of Agriculture and Fisheries only the responsibility for the translating and transferring any scientific advances to the farming community. It was especially important that the latter, especially as represented by the National Farmers Union, should be kept at a distance from all matters scientific because they viewed 'their opinions [as] purely local and based on purely local experience'.[20] Needless to say, the Ministry of Agriculture and Fisheries resisted. Farming needs had to be the main inspiration for research, ministers argued, and, in this respect, the Ministry of Agriculture and Fisheries was much better placed to assess those needs than was the Agricultural Research Council. The struggle was protracted, but the Agricultural Research Bill of 1956 eventually abolished the Development Commission and transferred control over all research funds to the Agricultural Research Council. It thus brought its institutional position closer in line with that of the Medical Research Council, but as a much more powerful arm of the corporatist state than might ever have been imagined.

The Plant Breeding Institute, the Scottish Plant Breeding Station, and the Welsh Plant Breeding Station were profoundly affected by this institutional transformation. Between 1945 and 1955, they were integrated into a centralised national agricultural research system. Their responsibility was to ensure that the first peacetime agricultural subsidies since the repeal of the Corn Laws would not be wasted on inefficient farming practices, such as sowing unproductive crop varieties. The accompanying funds from the Agricultural Research Council, vastly greater than they had ever received from the Development Commission, ensured that all three plant-breeding centres grew quite rapidly, if unevenly. The most remarkable expansion was that of the Plant Breeding Institute, probably because scientists trained in the University of Cambridge dominated the various committees of the Agricultural Research Council.

More importantly, however, the nature of work in the three plant-breeding centres changed uniformly. Up to this time, only a minority of the personnel at the Welsh Plant Breeding Station and Scottish Plant Breeding Station had identified with the Genetical Society, the main British association of geneticists. Now, they began to join the Genetical Society in greater numbers, as well as train or spend some time at the John Innes Horticultural Institution, the main British centre for training in genetics. Here, through the assimilating language of genetics, they learned how to link developments in Cyril Darlington's and J. B. S. Haldane's laboratories and the improvement of the very different crops for which they were responsible: wheat, barley, oats, forage grasses and potatoes. The once geographically, socially and culturally distant East Anglian fens, Welsh hills and the Scottish lowlands were thus brought into far closer proximity than was ever dreamt possible during the negotiations over the relationship between the Plant Breeding Institute and the Scottish Plant Breeding Station, for example. As Latour once put it, 'give me a laboratory and I will raise the world'.[21]

Strikingly, when the personnel of the three plant breeding centres published the results of their work in Cambridge, Edinburgh and Aberystwyth, their audience changed dramatically. It was neither the reader of publications addressing conceptual issues in genetics or other biological disciplines, nor the reader of *Farmer and Stockbreeder* or *Farmers' Weekly*, once the main vehicle for publicising the agricultural achievements of three plant-breeding centres. It was instead the reader of the *Annals of Applied Biology* or the *Journal of Agricultural Science*, that is, the reader of publications dedicated to problems peculiar to plant breeding, plant pathology, or other similarly agronomic sciences. In other words, the work of the three plant-breeding centres was becoming increasingly academic, but in a context where agricultural science was becoming a formal, if quite different academic discipline, deserving its own place within the walls of any British university, except perhaps in the University of Cambridge.

This transformation of plant-breeding research was strongly encouraged by the Agricultural Research Council on the understanding that research, untrammelled by economic considerations, was a necessary precondition for the advancement of its economic mission. This understanding was bolstered further as a generation of senior researchers returning from service in the now disbanding colonial empire argued that agricultural research, defined very widely, was the keystone of a productive national economy. It is then not surprising that many farmers, the new objects of imperial science, were said to regard the Agricultural Research Council as 'too theoretical and not sufficiently closely acquainted with the practical needs of farmers'.[22] Yet, if the researchers in the three plant-breeding centres were 'too theoretical and not sufficiently closely acquainted with the practical needs of farmers', their work was having an enormous impact in farmers' fields. With the introduction of subsidies, farmers were now willing to forgo their scepticism about investing in higher yielding varieties. Farmers' increasing attention to the seed they sowed was also intensified by contemporary changes in the structure of the food processing industries, such as the potato chips and then crisps industry, which called for larger procurements of more uniform raw materials. The uniformly and optimally shaped varieties of potatoes developed by the Scottish Plant Breeding Station began to displace a bewildering number of far more heterogeneous varieties that had been first produced during the mid-to-late nineteenth century. Similarly, oat and barley varieties produced by the Welsh Plant Breeding Station and the Plant Breeding Institute spread to occupy about half the national acreage dedicated to these two crops, which fed a burgeoning livestock and poultry industry. Even more impressively, during the early 1950s, Douglas Bell at the Plant Breeding Institute produced a new variety of barley, 'Proctor'. 'Proctor' proved so enormously popular that, by the end of the decade, it occupied almost 70 per cent of the national acreage, giving a new meaning to the term 'monoculture'. As Bell himself recognised quite explicitly, the success of 'Proctor' was due to two coincidental and unforeseen developments. The increased profitability of livestock and poultry fattened on barley meant that more barley was grown. This, in conjunction with the restrictions on imports imposed by the 'balance of payments' crisis, led to the substitution of the cheaper, but now larger and far more uniform domestic supplies of feeding barley for malting barley from Canada and the United States. A pint of Guinness may have begun to taste a little differently, but it was cheap and 'home-brewed'.

At the same time, the Agricultural Research Council, having identified a problematic reliance on imported seeds, turned aggressively to establishing the three plant-breeding centres as competitors with domestic seed firms. The latter were viewed as doing nothing helpful, even though major seed firms such as Gartons, Miln-Masters and Nickersons invested heavily in research facilities and jointly outspent the Agricultural Research Council. The seed firms, however, were dependent on sales, and it took ten to fifteen years of research and development before they could hope to realise a return on their investments. Since they were without the cushion that the Agricultural Research Council effectively provided for the plant-breeding centres, which were not expected to be financially dependent on the sale of their own varieties, these seed firms could ill-afford to be quite as adventurous as the three plant-breeding centres. Research into the mutagenic effects of colchicine, mustard gas and then radiation seemed too far removed from the business of breeding, even if it promised the production of novel plant types. These seed firms' situation only worsened with the introduction of legal protection of breeders' rights over new plant varieties. The directors of the three plant-breeding centres, especially Bell, the newest director of the Plant Breeding Institute, had doubted that the enactment of such protection was very important for the development of improved crop varieties. Nevertheless, they insisted that, if the proposed legislation were introduced, it should also cover the varieties they produced. In 1964, the directors' argument was accepted and, two years later, it was followed by the establishment of a National Seed Development Organisation, which would market and collect royalties from the sales of the varieties produced by their respective plant-breeding centres. Those firms that had once relied on the 'generosity' of the National Institute of Agricultural Botany, and had failed to maintain and expand their research facilities, now found themselves in a very difficult position. The same went for the more enterprising firms, since many of their varieties had incorporated lines developed by the three plant-breeding centres, for which they now had to pay royalties. In the meantime, funding from the Agricultural Research Council to the plant-breeding centres continued to increase ever more rapidly. 'Proctor' was repeatedly cited as the great example of what could be expected from the systematic application of advances of fundamental research produced by publicly funded science. Seed firms such as Nickersons were quite concerned about the likely results of competition with the National Seed Development Organisation. There were many discussions about their relationship to the National Seed Development Organisation, on which occasion they advocated the closer alliance adopted in many other countries and once pursued by the Welsh Plant Breeding Station and Dunns. Ultimately, however, the industry was too fractured to resist, and the by now re-named Ministry of Agriculture, Fisheries and Food decided that the entire process of research, development and production should rest in the public sector. Many seed firms quit the business of breeding. Nickersons turned toward Dutch breeders to maintain its position in the market. In 1978, however, the Anglo-Dutch petrochemical firm Shell acquired Nickersons, as it sought to bolster its growing investments in biotechnology by gaining a foothold in the increasingly very attractive British seed market.

Significantly, as I began this study of the institutional organisation of plant breeding in Britain, a representative of British Seed Company suggested that the firm once was very reliant on work at the Welsh Plant Breeding Station for the production of new grass varieties. By the 1970s, however, work at the Station focused so much on 'esoteric' questions that it could no longer be counted upon to help British Seed Company to maintain their commercial position.[23] Perhaps, however, this lack of support was the result of a tacit nationalisation of the seed industry, in the name of economic rationalisation and import substitution, rather than any explicitly ideological commitment: Conservative and Labour governments were equally supportive. Sir George Stapledon's dream of nationalisation had been quietly realised.

Privatisation and the coming of the 'age of genetic engineering'

During the 1960s, at the height of the corporatist and technocratic approach to the problems of the British economy, the institutional reorganisations following the passage of the Science and Technology Act removed the Ministry of Agriculture, Fisheries and Food even further from the management of agricultural research. As Prime Minister Harold Wilson put it in 1963, a new Britain was to be forged in the 'white heat of [the scientific] revolution'.[24] There then seemed to emerge a dual system of research. Some of the institutions under the control of the Ministry of Agriculture, Fisheries and Food, which were originally designated as centres for the more applied work outside the remit of the Agricultural Research Council, appeared to be engaged increasingly in areas of research formally assigned to the latter organisation. This was supposedly because they were not being investigated in a fashion sufficiently useful to advance the more practical concerns of the Ministry of Agriculture, Fisheries and Food.[25]

It was certainly the case that, by this time, the Plant Breeding Institute, the Scottish Plant Breeding Station, and the Welsh Plant

Breeding Station were subdivided into units separating 'basic' from 'applied' research. More specifically, some of these units were dedicated to research in 'cytogenetics', 'biochemistry' and 'developmental biology', and others were dedicated instead to 'plant breeding' and 'applied genetics'. Furthermore, the first set of units clearly enjoyed very close links with cognate departments in the associated Universities of Cambridge, Edinburgh and Wales, while the second set, always the smallest units, were more closely involved with economic problems. Yet, the understanding was that the separate units for basic and applied sciences would interact by being located in the same institution. They certainly did so very effectively, laying the ground from the late 1960s onward for the transformation of plant breeding into what Douglas Bell's successor, Sir Ralph Riley, began to call 'genetic engineering', which to my knowledge was the first use of the word in any British institution.

Nevertheless, in 1971, the so-called 'Rothschild Report' impelled a thorough reorganisation of agricultural research in response to increasing questions about the economic efficiency of public funding for research. Researchers in many institutions funded by the Agricultural Research Council misread the situation, and, by emphasising the economic nature of their work to protect their position, found themselves transferred to institutions controlled by the Ministry of Agriculture, Fisheries and Food. Yet, revealing its continuing allegiance to corporatist, technocratic principles, the Conservative government of the day established a 'customer–contractor' relationship between Agricultural Research Council and Ministry of Agriculture, Fisheries and Food. It did so notwithstanding critical voices that called into question the extent to which the Ministry of Agriculture, Fisheries and Food could in fact be viewed as speaking for the agricultural and increasingly powerful food industries. The net effect was then an even more thorough and wider ranging integration of academic research and corporatist and technocratic economic policy.

The relationship between the Agricultural Research Council and Ministry of Agriculture, Fisheries and Food was not in fact transformed in any profound way until the 1980s. A new and far more radical Conservative government rejected the bi-partisan, corporatist consensus, which had dominated policies for science, technology and the economy since the late 1930s. The by now renamed Agricultural and Food Research Council had been so effective in linking the worlds of academic science and the agricultural economy that it was hit particularly hard. The government proceeded to terminate or otherwise transfer into the private sector very large portions of agricultural research, development, and extension. Any institutions that were considered too small to be effectual were closed, and any staff not made 'redundant' were transferred to the remaining ones. The Scottish Plant Breeding Station was one victim of this policy. Those among the remaining institutions that were engaged in activities deemed to be the business of the private sector were instead driven to abandon such activities and become more involved in desirable kinds of research that were not likely to be funded privately. Thus, at a time of increasing concern over the 'environment', the Welsh Plant Breeding Station was transformed into the Agricultural and Food Research Council's Institute of Grassland and Environmental Research. Any institution that was even more heavily involved in areas of interest to the private sector and could be profitably privatised was privatised. The sale of the Plant Breeding Institute to Unilever, which beat off rival bids by ICI and Shell, was the most notable application of this principle. Ironically, however, the government miscalculated, and the profits from the privatisation reverted to the researchers once at the Plant Breeding Institute and now relocated in the new John Innes Centre at the University of East Anglia. Here these funds were used to build advanced facilities for biotechnology research in close association with the leading food processing industries in the nation. It should be no surprise that Lord Sainsbury, the chairman of Sainsbury PLC, should be a major patron of the University of East Anglia, as well as a minister for science and technology in the current Labour government who is particularly renowned for his championing of 'genetic engineering' and 'genetically modified foods'. Lastly, the legitimacy of public investment in agricultural research itself was called into question as a hangover of the corporatist state, so that the Agricultural and Food Research Council was disbanded and its remaining activities were transferred to a new Biotechnology and Biological Sciences Research Council. Genetics, the academic science that once lay at the heart of agricultural research, nominally for the benefit of the farmer, had finally been formally realigned with food processing industries, to transform Sir Rowland Biffen's wheat plant into a veritable biological machine. The old wheat varieties such as 'Squarehead Master', which Biffen had skilfully transformed by genetic tinkering into 'Yeoman', became 'Hereward', which is described by Plant Breeding International, now owned by Monsanto, in a fashion that conjures a technological artefact rather than a living organism:

> Hereward is a hard endosperm NABIM Group I variety which has found wide acceptance for breadmaking and is considered the best variety currently available for this purpose . . . Hereward has a standing score of '8', average tillering and a relatively slow rate of development. This combination allows a wide sowing window in which to achieve optimal yield – from mid September to mid November. Use of Roundup Biactive™ pre harvest can maintain Hagberg Falling Number and specific weight where changeable weather or harvesting workload risk [sic] delay in cutting.[26]

Echoing Louis Althusser, one might say that 'Frankenstein foods' had arrived, but more by 'overdetermination', than by any wilful and predetermined strategy. The logic of genetics, which transforms the phenomena of biological life into flows of genes ignorant of history and place, and logic of capital, which transforms the phenomena of political life into flows of commodities, which are equally ignorant of history and place, are homologous discursive formations.

As I articulated this otherwise satisfying narrative, however, I could not help but notice how the archive itself mirrored the increasingly impersonal organisation of the British state: names of people gave way to names of committees, as if these were in themselves veritable agents of history. The vibrant, polemical voices of the founding moments of this history disappeared. Perhaps hinting at my own alienation, I then found myself much more attracted to these founding moments, to which I returned to remember human agency.

Notes

1 See Segerstrale, *Defenders of the Truth.*
2 Anonymous, 'Seeds of dogma', p. 20.
3 This section is based partly on Robert Olby's 'Scientists and bureaucrats in the establishment of the John Innes Horticultural Institution under William Bateson'; and 'Social imperialism and state support for agricultural research in Edwardian Britain'.
4 Lloyd George, as quoted in Dale, *Daniel Hall*, p. 76.
5 Bateson, 'Toast of the Board of Agriculture', p. 76.
6 Parliamentary Debates, House of Lords, 14 April 1986.
7 Parliamentary Debates, House of Commons, 24 July 1986.
8 Tait and Urry, 'Monsanto pays £320m for UK crop breeding business', p. 33.
9 This and the following three sections are based partly on Palladino, 'The political economy of applied science'.
10 Wood, 'The School of Agriculture of the University of Cambridge', p. 229.

11 Archives of the John Innes Institute (Norwich): Plant Breeding Institute Collection: F. L. Engledow, *Plant Breeding: The Early Years*, p. 8.
12 Bell, 'The *Journal of Agricultural Science*, 1905-1980', p. 2.
13 Bell, 'Frank Leonard Engledow, 1890-1985', p. 200.
14 Hunter, 'Development in plant-breeding', p. 246.
15 Drummond, 'Report of the research director', p. 17.
16 This and the following section are based partly on De Jager, 'Pure science and practical interests'; Palladino, 'Science, technology, and the economy'; and Smith, 'The use of "team work" in the practical management of research in the inter-war period'.
17 Public Record Office (henceforth PRO): CAB 58/140, Economic Advisory Council, Report of the Committee on Agricultural Research Organisation, 1930, p. 4.
18 PRO: MAF 33/177, University of Cambridge, 11 May 1936.
19 PRO: CAB 58/148, Committee of Civil Research, Sub-committee on Agricultural Research Organisation, W. M. Fletcher, 21 January 1930.
20 PRO: MAF 33/756, Agricultural Research Council, Membership 1940-59, 17 April 1945.
21 Latour, 'The force and the reason of experiment', p. 76.
22 PRO: MAF 117/392, Agricultural Research Council, Working Party on Administration of Agricultural Research, 1954–55, 20 October 1954.
23 Neville Bark, personal communication, 15 December 1989.
24 Wilson, as quoted in Edgerton, 'The "white heat" revisited', p. 56.
25 This section is based partly on Thirtle, Palladino and Piesse, 'On the organisation of agricultural research in the United Kingdom'; and Webster, 'Privatisation of public sector research'.
26 www.pbi-camb.co.uk/cp_wheat_hereward.htm.

3

Genetics and the erasure of history

> There is a F4 family of 20,000 plants – not one (I believe) homozygous. How is that for orthodox genetics? Not much I think. Bateson is coming to see it soon but I wish you could come first to give me a few tips with which to comfort him. I read between the lines and although you write 1:2:1 here and there I am *sure* you think 1:2:1 is really for practical purposes rather futile.

Recovering agency

At the beginning of the twentieth century, many British botanists clearly believed that newly rediscovered mendelian principles of inheritance, which lie at the very heart of what we, following William Bateson, now call 'genetics', were about to transform the informal practices of plant breeding into an exact and precise science.[1] Echoing Galileo Galilei, plant breeding was to become a science of number and ratio. In 1911, a memorandum by the Board of Agriculture, outlining the programme of the Development Commission to promote agricultural research, suggested optimistically that there was 'a wide field for the application of mendelian principles in breeding new types of plants'.[2] Fifteen years later, the agricultural minister and head of Guinness Brewers, Walter Guinness, commissioned Victor Wilkins, a scientific officer in the Ministry of Agriculture and Fisheries and former graduate of the School of Agriculture in the University of Cambridge, to write a report on the organisation of agricultural research. In his report, *Research and the Land*, Wilkins wrote that,

> The actual methods of plant breeding are now fairly well established. Preliminary difficulties have been overcome, and for the most part the hybridising work is a matter of routine. The farmer appreciates, as a rule, the broad principles involved. He is becoming accustomed to the talk of Mendelism and inherited 'factors', and understands that by scientific methods of breeding it may be possible to combine in one variety several desirable qualities, and to introduce that variety to agriculture as an improvement.[3]

Significantly, Wilkins bolstered his claim to a conceptual closure around the relationship between genetics and plant breeding by citing the work of Sir Rowland Biffen in the Plant Breeding Institute.

In my earlier work on ecology and agricultural science, however, I had learned to be sceptical about academic scientists' claims about the transparence of agricultural practice. Thus I could not help but look for voices that would call into question these certainties about the relationship between plant breeding and the mendelian theory of heredity. Perhaps all too easily, I found that I did not have to dig too deeply into the archive of agricultural genetics to retrieve these dissonant voices. Although John Percival and Edwin Sloper Beaven were among Biffen's closest collaborators in forming both the British Seed Corn Association and the National Institute of Agricultural Botany, they often questioned the revolutionary importance Biffen and many of their contemporaries attached to the mendelian theory of heredity. I then proceeded to tell Percival's and Beaven's stories, as well as the story of the many, still more anonymous professional breeders who simply did not figure in John Jinks' and George Cooke's triumphal histories of genetics and agricultural research in Britain.

Plant breeding before the advent of the mendelian theory of heredity

According to many histories of genetics, before the advent of the mendelian theory of heredity, farmers sought to improve their crops by simply looking for that fraction of their crop that appeared to be more productive or otherwise attractive and then sowing part of it during the next growing season. This simple operation of 'mass selection' was repeated for a number of generations until farmers obtained plants that exhibited desirable characters to a high degree. Such mass selection may indeed have been the preferred method of breeding among farmers, but the practices of professional breeders employed by internationally renowned seed firms such as Gartons and Carter's Seeds were more complex, to say the least.

As early as in the 1880s, agricultural writers for the *Journal of the Royal Agricultural Society*, the *Journal of the Bath and West of England Agricultural Society*, or the *Transactions of the Highland and Agricultural Society of Scotland* reported that professional plant breeders preferred 'crossing' and 'hybridising' to mass selection. Apparently, these breeders collected varieties from all over the world and then crossed them to 'break the type'. They then selected from among the 'sports' thus produced the ones that exhibited the different characteristics that they wished to bring together in a single plant. Having sown the more interesting sports separately, they inbred the progeny of each one for a number of generations, always selecting those individuals that showed their peculiar characteristics in the higher degree. Once the desired character was 'fixed', that is to say, once they had a relatively stable line, in which the desired characteristics were uniformly transmitted from one generation to the next, the different lines were 'hybridised'. Each plant in the progeny of this last cross that exhibited the desired recombination was then inbred and screened until the 'composite cross' itself was fixed. This process could be repeated over and over to combine a number of desirable characters in a single plant.

Producing such 'composite crosses' was very time-consuming, especially since many of these crosses produced sterile progeny. Even in the simpler case of potatoes, which were usually reproduced vegetatively, or, to use contemporary terminology, by cloning, and thus 'departed from type' relatively rarely, it took about four or five years just to produce a useful new variety, let alone one that was commercially viable. Plant breeding was, therefore, an expensive process. As I explained in the last chapter, given the absence of any protection of proprietary rights over new varieties, smaller seed firms preferred to look for interesting varieties already on the market, and then appropriated them by multiplying and selling them under their own proprietary name. This motivated the support that organisations such as the British Seed Corn Association lent to the efforts to obtain state support for the Plant Breeding Institute, the Scottish Plant Breeding Station and the Welsh Plant Breeding Station. However, to move the Development Commission to support the creation of such institutions, the commercial breeders and academic botanists in the British Seed Corn Association had to argue that contemporary practices were somehow ineffectual.

Enter the mendelian theory of heredity

Between 1900 and 1905, Rowland Biffen, then a demonstrator in the School of Agriculture at the University of Cambridge, had been investigating how wheat varieties resistant to 'rusting', a common fungal disease, gradually lost their resistance. He did so with the encouragement of Henry Marshall Ward, a professor in the School of Botany, and William Bateson, a fellow of St John's College, who thought that Biffen's work might help to shed some light on the outstanding problem of the theory of evolution by natural selection.

For a number of years, Bateson and many of his contemporaries had been concerned with Charles Darwin's failure to explain exactly how new and advantageous traits spread and became the dominant features of a species, instead of being diluted by 'blending inheritance'. Although Francis Galton's name is today indelibly associated with the word 'eugenics', it is often forgotten how his experimental work on mechanics of inheritance aimed to resolve this very same problem. His theory that inheritance was mediated by blood-borne 'gemmules' was not convincing, however. In 1900, Bateson famously rediscovered the work of Gregor Mendel, which immediately struck him as offering a better answer to the problem by positing that the heritable characteristics of an organism were determined by the recombination of discrete, germinal 'mendelian factors'. On the other hand, he and a number of his contemporaries worried that these mendelian factors were no more than useful accounting devices to order the multiplicity of forms produced by the crossing of two individuals. It was in this context that Bateson encouraged Biffen to examine fungal resistance from a mendelian perspective.

Biffen began repeatedly to self-fertilise individual plants of wheat and then segregate their progeny by type. This eventually produced a number of distinct 'pure lines'. That is, some of these lines were uniformly resistant to rusting, and some were instead uniformly susceptible to rusting. More importantly, once these lines had been thus purified, Biffen demonstrated that they were unalterable. He concluded therefore that the susceptibility of these lines to rusting was fixed by a mendelian factor, and no amount of exposure to fungal spores would change this, except perhaps for extremely rare 'germinal' sports, which today we would call 'genetic mutants'. On this understanding, the declining resistance to rusting, which had been observed in farmers' fields, was due to farmers' habit of sowing that fraction of the crop that was more resistant to rusting without paying any attention to individual variability and worth. In other words, resistance had become a fundamentally individualised characteristic. Consequently, any improvement that the farmers might obtain by 'mass selection' could only be a temporary phenomenon. It would quickly be lost as interbreeding among the genetically heterogeneous progeny of this fraction scrambled the resistant combination upon which the farmers had accidentally and unknowingly stumbled.

Among a growing number of botanists and zoologists who were identifying themselves increasingly as 'geneticists', Biffen's work was a fundamentally important contribution to their new discipline. It demonstrated that the mendelian principles of heredity were applicable not just to simple morphological characters, such as the smoothness and wrinkling of Mendel's celebrated peas, but also applied to complex physiological characteristics such as rusting. More importantly, perhaps, Biffen's work also went some way toward demonstrating that mendelian factors were not simply useful heuristic fictions, but veritable biological entities. Although the problems besetting the theory of evolution by natural selection were not yet resolved, this provided these botanists and zoologists with the wherewithal to address the problems experimentally.[4]

Biffen, however, was interested in doing much more than simply bolstering the credentials of genetics among botanists and zoologists interested in the mechanics of the theory of evolution by natural selection. He argued that, if farmers wanted to reduce their losses to rusting, they had to sow specially designed seed. Combining his systematically isolated, and genetically pure, lines of wheat with other equally systematically isolated lines that displayed other desirable characteristics such as high yield could produce such seed. His particular approach to hybridisation, moreover, was superior to that of Gartons or Carter's Seeds because, he claimed, its outcome was mathematically predictable. Given knowledge of the number of the discrete, heritable factors controlling the different characteristics, it was possible to identify which of a number of seemingly identical progeny of a cross would display the desired combination in an unalterable fashion once sown in the farmer's field.

By 1910, with much support from A. E. Humphries, the secretary of the Home-Grown Wheat Association, Biffen demonstrated the viability of his mendelian approach to plant breeding. He hybridised a line of 'Squarehead Master' that was particularly high yielding, but susceptible to rust, with a low yielding but resistant line of 'Girka', to produce a new variety, 'Little Joss', which incorporated the two highly desirable characteristics.

The British Seed Corn Association and the Home-Grown Wheat Association soon translated the popularity of 'Little Joss' among farmers around East Anglia into an outstanding example of how academic investigations could transform agriculture. Within two years, their campaign to move the Development Commission to support the creation of institutions dedicated especially to plant breeding resulted in the establishment of the Plant Breeding Institute.

In the meantime, Biffen turned his attention to other problems of heredity in wheat. In particular, he turned to the so-called 'strength' of wheat, a characteristic of the grain that was of the utmost importance to flour millers. Bread makers were dissatisfied with the flour extracted from British wheat, preferring instead the stronger flour that could be extracted from Canadian and American wheat. In 1907, Biffen had claimed quite controversially that such strength was inherited in mendelian fashion. It then seemed to him obvious that a strong line could be combined with a high yielding one that was suitable for cultivation in Britain. In the wake of Biffen's success in producing a variety resistant to rusting, Humphries and the Home-Grown Wheat Association lent Biffen their fullest support. Within ten years, and despite the considerable scepticism among, and criticism of, other breeders and academic botanists, Biffen produced a new wheat variety, 'Yeoman', which was significantly stronger than those then cultivated by British farmers.

For Biffen and a growing number of advocates of the agricultural research institutions supported by the Development Commission, 'Yeoman' was an even more exemplary triumph of the application of academic research to the solution of very important agricultural problems. It was not just academic researchers who believed so. Developing 'Yeoman' earned Biffen the accolades of many farmers' clubs throughout Britain. Of course, the development of 'Yeoman' also spurred further the creation of the National Institute of Agricultural Botany, since seed firms were not prepared to countenance that the income from sowing upward of 10 per cent of the national wheat acreage with 'Yeoman' should go to the Plant Breeding Institute. While Biffen was rewarded with a knighthood for his service to agriculture, the art of plant breeding was transformed into the scientific enterprise of 'applied genetics', securely located in an institutional network mediating between the public and private sectors.

An academic disagreement?

At the same time, however, John Percival, in the Department of Agriculture at the University College of Reading, was profoundly sceptical about Sir Rowland Biffen's achievements. Percival agreed with most of his contemporaries that the mendelian principles of inheritance provided greater understanding of the bewildering number of variants produced by crossing. He also argued, however, that the pivotal mendelian assumption of the independent segregation of characters was applicable only to morphological characteristics of the wheat plant. Those characteristics that most interested farmers, yield and strength, depended on such complex physiological and environmental interactions that they could not be regarded in the same fashion. Metabolic pathways could indeed be shunted by selection and inbreeding toward the production of more starch, one of the factors underlying yield, but only at the cost of interfering with the process of nitrogen assimilation, one of the chief factors thought to be responsible for strength. Moreover, however heritable, the expression of these characteristics was intimately connected to environmental factors such as climate and soil fertility. Current knowledge of these interactions was simply inadequate to create pure lines that could be used, like chemists' molecules, to precisely construct a plant that met farmers' requirements. Furthermore, crossing was just as likely to be disruptive as constructive. As Percival pointed out, commercial breeders used hybridisation to 'break the type', and only a limited number of the variants thus created were viable. The wheat plant was a very complex organism, finely attuned to its environment by a long history of evolutionary adaptation.

Percival's most trenchant criticism was, however, that the notion of a genetically uniform pure line that was refractory to further selection was not applicable outside Biffen's experimental plots. What Biffen called 'germinal' sports might indeed be extremely rare in such plots, but in the farmer's field individuals belonging to the same line were counted in the millions and such sports were likely to be far more common. Hence, it made sense for farmers to continue looking for exceptional plants in their crops, even if they had sown seed developed along the lines pursued by the Plant Breeding Institute. For commercial and professional breeders, the more effective approach was instead to gain a more precise understanding of the multiplicity of already existent types, and, in particular, a more precise understanding of their historical adaptation to particular climatic and soil conditions. It was in just this fashion that Percival provided farmers around Reading with more productive 'selections' of 'Cone', an old and established variety of wheat.

It is perhaps not surprising then that Percival's *The Wheat Plant*, a catalogue of wheat strains being cultivated around the world, detailing their genealogical relationships, was both summarily dismissed by Biffen and praised fulsomely by the Russian bio-geographer and pioneering evolutionary geneticist Nikolai Vavilov. For Vavilov, as for Percival, place and history were fundamentally important for understanding the evolution of new species.

Significantly, Percival sometimes shared his scepticism about pure lines with Edwin Sloper Beaven, the foremost professional plant breeder in Britain. Thus, when faced with a visit from William Bateson, Beaven wrote to Percival:

> There is a F4 family of 20,000 plants – not one (I believe) homozygous. How is that for orthodox genetics? Not much I think. Bateson is coming to see it soon but I wish you could come first to give me a few tips with which to comfort him. I read between the lines and although you write 1:2:1 here and there I am *sure* you think 1:2:1 is really for practical purposes rather futile. (Emphasis in the original).[5]

Like Percival, Beaven damned the mendelian theory of heredity with faint praise. He argued that the theory had contributed greatly to the improvement of plant breeding. It had encouraged breeders to think of plants as composites of separate characteristics that could be recombined, and to be more systematic about the selection of individuals resulting from hybridisation. Beaven also argued, however, that the mendelian theory reduced the complexity of inheritance too far.

Beaven had embarked on his career as a plant breeder after studying the effects of chemical fertilisers on the malting quality of barley, concluding that they could not be addressed properly without considering which variety of barley was sown. Different varieties responded differently. He had hoped to exploit this differential response by following the practices pioneered by Biffen, discovering, however, that, even in pure lines, the malting quality of barley varied enormously under different regimes of soil fertility. More importantly, he also found that selection within these pure lines could change their malting quality. He concluded that pure lines were not in fact genetically homogenous, perhaps because malting quality was determined by such a great number of mendelian factors that breeding a genetically pure line whose malting quality was absolutely uniform and unchanging was impossible. Alternatively, he speculated, the environment could alter the mendelian factors. In 1911, Beaven discussed this last matter with his close friend and famed evolutionist Alfred Russel Wallace, suggesting that acquired characteristics might be inherited. Even though one might presume that Wallace was not sympathetic to this increasingly heretical idea, Beaven must not have been sufficiently reassured, for even as late as in the 1930s he remained 'sceptical with regard to the non-transmission of characters varying genetically in response to the environment'.[6] Still, Beaven never committed himself in any more explicit way to the neo-lamarckian cause, and usually spoke rather ambiguously of the environment as the most important factor determining the characteristics of a plant.

On the basis of these considerations about the importance of environmental factors in shaping the phenomena of heredity, Beaven argued that the most effective approach to breeding was to select the one plant best adapted to the environment in which it was to grow. This was presumably done on the basis of historical experience. It was of course possible that two well-adapted plants might exhibit different and desirable characters. If so, following the 'purely empirical practice of all successful breeders', the best plants in these populations should then be crossed to obtain a hybrid.[7] It was in just this manner that, between 1900 and 1920, Beaven produced some of the most popular barley varieties produced before the advent of Douglas Bell's spectacularly successful 'Proctor'. Beaven's accompanying fame was not limited solely to Britain, but spread across the Atlantic. Influential American agricultural researchers, such as Albert Mann of Cornell University, visited Beaven's 'research station', or 'garden', as Beaven preferred to call it. Mann was intent on collecting information about the organisation of agricultural research for the United States Department of Agriculture, then in the throes of radical transformation, and was very favourably impressed by Beaven's views on plant breeding. The Development Commission apparently thought otherwise, when it refused Beaven's applications for funding.

The making of breeders

As I mentioned in the opening of the last chapter, one of the most exciting features of coming to Britain to work on the history of plant breeding in this country was to further my scant knowledge of social constructionism. I then turned to Sir Rowland Biffen, John Percival, and Edwin Sloper Beaven's biographies to explain sociologically their radically different views of the relationship between the mendelian theory of heredity and plant breeding.

As I scoured through minor archives across Britain, corresponded with surviving relatives, and immersed myself in the literature on the professionalisation of science, I 'discovered' that Biffen's career was a paradigmatic example of the making of an agricultural researcher in University of Cambridge. As the talented son of a headmaster in a minor public school, it was almost expected that Biffen should study in the University of Cambridge. In 1895, after sitting the exams in natural history and obtaining a 'double-first', he was awarded a Smart Scholarship in Botany. During the next few years he studied mycology, the science of fungi, under the guidance of Henry Marshall Ward, who introduced Biffen to the 'New Botany'. This experimentalist school of botany sought to displace the older tradition of natural history and taxonomic ordering, and to gain greater visibility and financial support for its programme by undertaking 'economic' work. Thus, in 1898, when Biffen was appointed demonstrator in the School of Botany, he followed Marshall Ward's example and began to study the interaction of pathogenic fungi and their hosts in an economically significant plant. He turned his attention to the 'rusting' of wheat, which attracted the support of the Home-Grown Wheat Association.

Percival's career, which culminated with a professorship in agriculture at the University College of Reading, was very different. He was born into the family of a small farmer in rural Wensleydale and, after leaving the local village school, went on to work in a glass-making factory. In 1884, Percival was able to enter the University of Cambridge thanks to the generosity of the owner of the glassworks. Like Biffen, he sat the natural sciences examinations and went on to study plant pathology under Marshall Ward. After finishing his studies, he moved to London, where he worked in the Department of Botany at the British Museum of Natural History. In 1891, Percival returned to the University of Cambridge as a demonstrator in chemistry and there joined Daniel Hall as agricultural extension lecturer. At this time, Hall was the leading figure in the movement to establish agricultural education as an important component of the scheme for the promotion of technical education among the industrial and agricultural working classes, which had been recently inaugurated by the then still new County Councils. Over the next three years, Percival lectured farmers and growers on the agricultural applications of chemistry and botany. Three years later he followed Hall to Wye, where he was appointed professor of botany in the newly established South-

Eastern Agricultural College. While at Wye, Percival wrote the first, and immensely popular, textbook of agricultural botany published in Britain. Its aim was explicitly the preparation of students intending to obtain the National Diploma of Agriculture, rather than the degree offered by the School of Agriculture in the University of Cambridge. Percival also began to turn his skills in plant physiology and systematic botany toward the study of the wheat plant. The Home-Grown Wheat Association aided this effort, just as it was doing with respect to Biffen. In 1902, Hall and Percival parted ways. Hall left Wye to become the director of Sir John Lawes and Sir Henry Gilbert's world-renowned experimental station in Rothamsted, while Percival moved to the University College in Reading as professor of agriculture and director of the Agricultural Department.

While at Reading, where he brought his academic wandering to an end, Percival continued to be actively involved in the agricultural education and extension movement. In the meantime, he was also involved in the transformation of the Agricultural Education Association. This national association of educators interested in agriculture was established in 1894. Initially, academics from the University of Cambridge dominated the Association, but during the following decade, as they established a more secure institutional situation in their university, they gave way to academics from other institutions, especially from the University College in Reading. In 1903, the members of the Association rejected the proposal by academics from the University of Cambridge to impose a levy on their membership fee for the purpose of establishing a fund for agricultural research. The members admitted, however, that the nature of their organisation was changing by agreeing to alter the name of the Agricultural Education Association to the Association of Economic Biologists, but their different understanding of the nature of agricultural research was captured by their new publication. The academics in the University of Cambridge busied themselves organising the *Journal of Agricultural Science*, which, as I have already noted, would not accept papers dealing with 'farming as opposed to agricultural science'.[8] Biffen's work on rusting and the mendelian theory of heredity appeared in the very first volume of the *Journal of Agricultural Science.* The Association of Economic Biologists instead established the *Journal of Economic Biology*, on whose editorial board Percival often sat, as an alternative medium of communication for those researchers whose work would not be published by the *Journal of Agricultural Science.* In sum, Percival's approach to agriculture was rather more orientated toward economic matters than was Biffen's.

Beaven's career was completely different from either Biffen's or Percival's. He was born into a family of prosperous Wiltshire farmers, who sent him to be educated in a minor public school. There he may have become first interested in science thanks to the attention bestowed upon him by his tutor in chemistry. However, at the age of thirteen, he was removed from the school. Apparently, his father had little taste for this tutor's view that he should not be teaching his son 'facts', but how to find out for himself. All that Beaven took with him from his school was a Certificate of Proficiency in Religious Knowledge, which, in later life, would serve him well to lambast Friedrich Nietzsche's 'nihilism' during a Sunday School sermon.[9] In the meantime, he began to learn the business of farming on his father's farm. He continued working on the farm after his mother took over its management upon the death of her husband in 1872. However, six years later, in the depths of the agricultural depression, the farm was sold off. Beaven was then driven to take the first steps in his career in the malting industries by joining Morgan Maltings of Warminster. Having married Mr Morgan's daughter, he eventually became the owner of Morgan Maltings.

During the early 1880s, motivated by the variable malting quality of barley available on the British market, Beaven, now a professional maltster, became interested in the factors affecting this variability. He began corresponding with the professor of botany and breeder in the agricultural college at Bonn-Poppelsdorf, Friedrich Körnicke, and then with the Danish botanist Carl Hansen, who, at the time, was working in close association with Carlsberg, the Danish brewers. Beaven obtained from Körnicke and Hansen a vast collection of barley varieties, which he began to grow in his 'garden' at Warminster, aiming to understand how their differences were affected by the conditions of cultivation. By 1894, Beaven had become sufficiently wealthy to afford the cost of hiring two assistants to conduct these botanical experiments on an increasingly larger scale. He also exchanged the results of these experiments with academics associated with the Wiltshire Research Committee, one of the many organisations established by county councils to support the local agricultural economies. One of these academics, John Munro, an agricultural instructor at the Downton College of Agriculture, was also collaborating in Lawes and Gilbert's own studies in Rothamsted. Like Beaven, Lawes and Gilbert were interested in understanding how environmental factors, especially the artificial fertilisers that made their fame and fortune, affected the productivity and malting quality of barley and wheat. Lawes and Gilbert were very impressed by Beaven's work. They then called on his assistance to prepare a report for the Parliamentary Committee on Beer Materials, on the quantities of barley produced at home and imported from abroad, on the origin of the imported barley, and on the malting quality of the different varieties. These inquiries brought Beaven into contact with a number of scientists working in one function or another for Guinness Brewers, chiefly the plant breeders Alan McMullen and Herbert Hunter, the chemist Horace Browne, and the famous statistician

William 'Student' Gossett. By 1900, Beaven was malting on a contract basis for Guinness Brewers, and, through Browne and Gossett, was involved in the work of agricultural researchers in the School of Agriculture at the University of Cambridge. Browne introduced him to Marshall Ward, who was very interested in Beaven's work on varietal differences and drew his attention to the mendelian theory of heredity, which was then being studied by Ward's former student, Biffen. With these associations began the most productive period in Beaven's 'scientific' career.

Academics, professionals and the politics of science

In social constructionist fashion, I then went on to argue that Sir Rowland Biffen, John Percival and Edwin Sloper Beaven's very different social backgrounds and careers shaped their views on the relationship between science and agriculture. More specifically, they shaped quite profoundly their views on the relationship between the mendelian theory of heredity and plant breeding.

As Henry Marshall Ward's programme for the development of the 'New Botany' illustrates, scientists in the University of Cambridge were not averse to becoming involved in economic activities, but doing so was fraught with difficulties, chiefly overcoming the fear that this would taint the intellectual standing of the University. Resistance to the establishment of the School of Agriculture was a prime example of such difficulty. One answer was to argue that scientific research was the prime mover of economic change, but only when absolutely freed from any partisan motives. Thus, if representatives of the agricultural industries were not to be allowed to undermine the autonomy of botany, geology, or zoology, by dictating the direction of scientific research, the same went for agricultural education. Understanding the improvement of agriculture as a matter of educating farmers to adopt more advanced methods of production would be just as distracting as any input from the agricultural industries. The improvement of agriculture had to be understood instead a matter of scientific investigation by academic scientists who were trained in traditional academic disciplines and would eventually turn the products of their labours over to the farmers or other members of the agricultural community for application in the field. This philosophy was eventually formalised by the Committee on the Neglect of Science, which was convened during the First World War to both encourage greater government support for science in British universities and protect the traditional academic values cherished by academics in the University of Cambridge. Biffen and the Plant Breeding Institute provided outstanding evidence of what could be achieved in this manner. It should not be surprising that Biffen was a member of the Committee on the Neglect of Science and a contributor to its published report, Sir Albert Seward's *Science and the Nation.*

Establishing the causal dependence of economically effective plant breeding on the systematic application of an autonomously developed and institutionally removed mendelian theory of heredity certainly required considerable rhetorical dexterity. Yet, if the fuller articulation of this theory was to have the impact promised by the authors of *Science and the Nation,* it had to be linked to economic activity in a more socially concrete fashion than the philosophy articulated in *Science and the Nation* would ever allow. This took a peculiar form in the University of Cambridge. The School of Agriculture maintained a direct connection with the world of agriculture by recruiting the most promising students into its teaching ranks. Moreover, these students were not the sons of any farmers, but of large landowners closely linked to 'progressive' organisations such as the Home-Grown Wheat Association. Significantly, it was thanks to A. E. Humphries, the secretary of the Association, that Biffen came to believe that developing stronger wheat was the most appropriate way of meeting the needs of agriculture. Humphries and Biffen's productive encounter, however, was far from accidental because Humphries was one of the many graduates of the School of Agriculture, who had learned from Biffen and others that agriculture was an experimental practice in maximising the output of crops and farm animals. The effectiveness of this practice depended on a thorough understanding of underlying biological principles such as the mendelian theory of heredity, even if this meant devoting time to learning all there was to know about the biology of the ornamental plants favoured by William Bateson and his collaborators. After all, the wheat plant was a plant like any other encountered in the world of botany, and the plants of 'Little Joss' Biffen nurtured in the tidy experimental plots of the Plant Breeding Institute testified to this fundamental similarity. Needless to say, much more could be expected of this fundamental similarity if others in the Home-Grown Wheat Association, beside Humphries, came to understand as much. In sum, Biffen and Humphries' extended engagement bolstered the significance of the mendelian theory of heredity far beyond anything that the philosophy articulated by *Science and the Nation* could ever accomplish.

Percival, on the other hand, would not conflate and elide the differences between biology and agriculture as readily as Biffen seemed prepared to do. Thus, he objected strenuously to the emphasis the Home-Grown Wheat Association placed on improving the strength of the wheat because he believed that such improvement could only be obtained by decreasing yield. Given that the millers represented by the Home-Grown Wheat Association were not prepared to pay farmers a higher price to compensate for the lost yield, growing stronger wheat was not an economic proposition. Percival then argued that what farmers needed above all was higher yielding varieties. This issue, of course, brought him into conflict with Biffen, who could not conceive that the interests of millers and farmers could possibly diverge and believed anyway that the fuller articulation of the mendelian theory of heredity would deliver both stronger and higher yielding wheat varieties. Significantly, when challenged by farmers on the relationship between millers and farmers, Biffen responded very revealingly that he knew 'nothing about farming'.[10] Far more importantly, Percival argued that the promise of the mendelian theory of heredity was overblown, and that selection could yield still greater improvement if farmers could be taught how to assess more precisely their needs and how to assess the merits of the sports they found in their fields. This would allow them to develop varieties more suited to their specific needs. For Percival, in other words, the farmers themselves had to become the agents of scientific improvement in agriculture.

Percival's radically different perspective was perhaps inevitable for a member of the University College in Reading. The College was established in 1892 to meet the needs of the local farming and horticultural community, which could not depend on nearby Oxford University for technical education and advisory services. The Palmer family, founder of the biscuit-making company Huntley and Palmers of Reading, was the driving force behind the organisation and orientation of the college as a focus of civic activity and corporate identity. The Sutton family, owner of Sutton's Seeds and major supporter of William Bateson's pioneering genetic investigations, and Lord Wantage, a landowner and stockbreeder of considerable wealth, contributed to this effort just as much. Moreover, Alfred Palmer and Leonard Sutton sought to encourage the modelling of the College after the American land-grant universities. The academic community, however, resisted their efforts quite successfully. They rejected the proposal that laymen should have a voice in professorial and educational matters, as was reputedly the case in the American universities. Thus, although more inclined toward meeting the needs of local farmers and horticulturalists, the orientation of courses provided by the staff of the Faculty of Agriculture was largely defined by the academics themselves. Nevertheless, the atmosphere at Reading was fundamentally different from that in University of Cambridge, as it had to be to sustain recruitment. Thus, although students in Cambridge could obtain their degree in agriculture without any previous experience of farming or horticulture, such experience was a prerequisite for students seeking admission to the course in Reading. Even after being admitted, these students spent considerably more time than their counterparts in Cambridge working in fields, orchards, and farm-yards. Furthermore, a great deal of attention was paid by the faculty to integrating course schedules and the timetable dictated by work on the farm and orchard so that part-time enrolment might be a practical option for their agricultural students. Strikingly, many of these students were women, who hoped to improve their ever more difficult prospects of a livelihood in agriculture by obtaining a National Diploma or a university degree. This orientation toward meeting the needs of farmers and horticulturalists was so firmly entrenched at Reading that in 1926 only six of nearly three hundred graduates in agriculture, horticulture, or dairying earned a university degree; the others either attended short courses or obtained National Diplomas. Even as late as 1938, students earning the National Diplomas constituted half of the student body taught by the staff in the Faculty of Agriculture. During the same year, the University of Cambridge granted over a hundred three-year undergraduate degrees and thirty-five postgraduate diplomas in agriculture. The aim of the agricultural education at Reading was not to create a body of experts who could solve the scientific problems confronting the farmer or horticultural producer. It was instead to improve farmers' and horticulturalists' own understanding of the scientific foundations of their practical activities so that they might improve the productivity of their holdings.

As with his career, Beaven's understanding of the relationship between biology and agriculture was completely different from either Biffen's or Percival's. He was first and foremost a businessman. Thus, when requested by Sir Henry Gilbert to contribute to the report to the Parliamentary Committee on Beer Materials any information he might have on the barley varieties to be found on the British market, he declined. His reason was simply that 'it is much easier to do oneself harm than good from trading points of view'.[11] For Beaven, the search for scientific knowledge was a financial investment, which should be treated as such. The preface to his autobiography captures the profoundly utilitarian spirit of his work. He argued, for example, that,

> What is really important is that the search for new knowledge and its practical application to industry, or purpose, to which that knowledge is allied should progress, should be welcomed, and should receive its due reward.[12]

He also wrote that,

> The primary object of all these investigations and discoveries was not the increase of pure knowledge but very definitely the acquisition of knowledge which could be put to practical use.[13]

Furthermore, like Robert Kargon's good Victorian 'devotee' of science, Beaven thought that the general improvement of farming conditions could not possibly rest on any action of government, but rather should be the responsibility of landowners and farmers themselves. He often liked to argue that the most remarkable achievements in agricultural improvement were the work of self-motivated men of his own class. Ironically, he seemed to consider his work as far more methodical than that of his then most famous academic colleague, Biffen. He believed that plant breeding required both 'art' and 'science', and that Biffen excelled in both and for that reason was justly called the 'wheat wizard'. He, on the other hand, was no artist, and had to depend therefore on 'more laborious and meticulous methods'.[14] It is then not surprising that he looked upon those academics who called for state support for science with disdain, if not suspicion. It was an investment in 'wizardry'. Nevertheless, he believed in the need for some minimal governmental funding for scientific work. If it could not be channelled directly to more pragmatic and economically motivated people like himself, then it should only go to institutions closely supervised by bodies such as the British Seed Corn Association, which he himself helped to establish. Not surprisingly, and unhappily for many of his friends in the University of Cambridge, Beaven became a vocal and constant critic of the quite differently orientated network of agricultural research institutes which was being built in Britain from the beginning of the century. Indeed, at the end of his life, he claimed to be thankful that his own research had not been 'subsidised', for government intervention could only corrupt scientific inquiry.[15] This said, Beaven, craved the approval of researchers in academic research institutions. As I pointed out earlier, he collaborated quite extensively with leading researchers at the University of Cambridge. In return, when the university decided in 1922 to grant six honorary degrees for signal contributions to agricultural improvement, Beaven was one of the recipients. These researchers also nominated Beaven for a fellowship of the Royal Society, whose rejection he bitterly resented. In the fashion of a Victorian novel, however, Beaven at least succeeded in marrying his daughter and sometimes assistant, Alice, to the Professor of Agriculture, Thomas Barlow Wood.

In sum, during the twenty years after the rediscovery of Gregor Mendel's work, there was considerable disagreement among British breeders, botanists and geneticists about the nature of pre-mendelian practices of plant breeding, as about the importance of the theory to post-mendelian practice. Biffen, Percival, and Beaven were major actors in this dispute. The first two had both been introduced to the 'New Botany' under the guidance of Marshall Ward while students in the University of Cambridge, and might then have then been expected to agree on the importance of the mendelian theory of heredity. Biffen and Percival, however, disagreed, on the relationship between this theory and the practice of plant breeding by drawing on their common training in a way that reflected different conceptions of the relationship between science and the farmer. These conceptions were characteristic of the universities at which they worked, namely the University of Cambridge and the University College in Reading. Beaven's criticism of the mendelian theory of heredity, on the other hand, reflected the utilitarian values of the self-made businessman that he was, and the accompanying, profound scepticism about academics and the funding they received from the state. Thus, while writing about his work to Alfred Russel Wallace, who was also known to many of his contemporaries as a socialist thinker, Beaven protested against the slide of his beloved country into 'socialism' and swore his continuing allegiance to the social theories of his other old friend, Herbert Spencer.[16]

Alliances and forgetting

This account of the differences between Sir Rowland Biffen, John Percival and Edwin Sloper Beaven does not explain, however, how Biffen's controversial approach to plant breeding came to dominate the organisation of agricultural research from the 1930s onward, and, ultimately, paved the way for Monsanto's extraordinary investment in Plant Breeding International.

Leading figures in the Development Commission, such as Daniel Hall, were clearly aware of Beaven's critical views on the institutional organisation of agricultural science. In 1913, just as the Plant Breeding Institute was being established, Hall dedicated his *Pilgrimage of British Farming*, a review agricultural practices across the country, to Beaven, by writing that,

> I am always reminded, sometimes of your methods of dealing with barley problems, the rigour of which may well put us professional men of science to shame, sometimes of the acuity of your criticisms of the work of the same class of pundits. But most of all I think of you as one of the last defenders of the old laissez faire position, a latter-day Athanasius, standing for self-help and honest individual work, and denouncing Government Departments, County Councils, Development Commissions, and all such spoon-feeding agencies, until that ancient thruster – the old Leon Bollee, would boil over with the combined fury of your driving and your arguments! May it be granted to us to review our opinions, years hence, over the same ground![17]

Hall's hopes that Beaven might eventually review his 'opinions' were not fulfilled. Even as late as in 1938, this last champion of 'laissez faire' still managed to worry figures such as E. J. Butler, the secretary of the Agricultural Research Council. During a public discussion in the Farmer's Club, Butler admitted that,

> The scientific man in this country has often a very hard row to hoe, because the amateur breeder, the non-research-institute breeder, has made such remarkable progress that the scientific man does not find much room for direct improvement; that is due to the work of men like Dr. Beaven himself.[18]

Beaven was in no conciliatory mood. Three years after the discussion in the Farmer's Club, he wrote that 'agricultural research is a task for practical rather than for (strictly speaking) scientific men'.[19]

Despite such competition, however, Beaven and administrators such as Hall and Butler agreed that the future of agriculture rested with corporate organisations such as the Home-Grown Wheat Association or Guinness Brewers, rather than the farmers championed by Percival. Sometimes this meant setting aside arguments over the relationship between academic science and agricultural practice. When Biffen nominated Herbert Hunter as his successor at the helm of the Plant Breeding Institute, Hall, by then Sir Daniel Hall, doubted the propriety of the nomination. As I mentioned in the last chapter, Hunter was for him 'a plant breeder and not a geneticist [who] has little interest in the important fundamental basis of a plant-breeding programme'.[20] Hunter, however, was closely connected with Guinness Brewers, and as such, his appointment promised to cement what were by this time strong links between the Plant Breeding Institute and one of the most influential processors of agricultural products anywhere in the British Empire. Beaven and Hunter's reservations about the relationship between academic science and agricultural practice could be overcome anyway by deft publicists such as Sir Walter Morley Fletcher, the champion of the scientific transformation of both medicine and agriculture. In one particular radio broadcast in 1935, Fletcher argued that Beaven was a champion of agricultural science and fully supportive of the Agricultural Research Council. Even more outrageously, Beaven's obituary in *The Times* suggested that '[Beaven] may be looked upon as the founder of the modern school of thought, working on the mendelian theory of inheritance . . .'.[21]

Clearly, geneticists were hitching their wagon to a renowned professional plant breeder, but in a way that Beaven would have certainly rejected. He was dead, however, and the geneticists got to write the history of the relationship between genetics and plant breeding. No professional plant breeder sought to write an alternative history. Even Beaven's polemical autobiography was published posthumously. Percival and Beaven's dissonant voices thus disappeared into the past. In sum, the historicity of plants was broken in more than one way.

This said, over ten years ago, spurred by Stephen Jay Gould and Richard Lewontin's critiques of genetics, which have sometimes been associated with a critique of capitalism, I began to rummage through archives dispersed around Britain and sought surviving relatives to return John Percival and Edwin Sloper Beaven to history. I did so with the intention of telling a 'history from below', by deploying the conceptual framework of social constructionism, hoping to thus challenge the hegemony of genetics and its accompanying, all too invisible political formations. This alliance was so powerful that it compelled even the most critically aware friends and colleagues to wonder about my interest in the most 'obscure historical figures'. All that I in fact succeeded in accomplishing in the intervening years, however, was to recover Percival and Beaven's voices, to then re-imprison them in Max Weber's famous 'iron cage'. I had reduced them both to figures in the institutional landscape of 'modernity', that is, the historical realisation of the Enlightenment which has resulted in a landscape utterly devoid of any agency except that of the state and capital.

Strikingly, I was no freer than Percival or Beaven, as I was soon forced to switch from the history of agricultural genetics to that of medical genetics. I intended, however, to explore the uncanny complicity of social constructionism, state, and capital to which Michel Foucault's *The Order of Things* seemed to speak. Originally, I had turned to *The Order of Things* simply to understand the significance that Percival and Montagu Drummond, the first director of the Scottish Plant Breeding Station, attached to taxonomy and natural history as counters to the mendelian theory of heredity. As I immersed myself in *The Order of Things*, however, I became increasingly aware that, for Foucault, the emergence of 'human sciences' such as sociology was the manifestation of a new discursive formation. Foucault clearly associated this development with the emergence of the modern state. Moreover, as Michael Hardt and Antonio Negri have noted, Foucault sometimes also associated it with the history of capital, albeit in a way that was inadequate to the truth of 'life' as such.

Notes

1 This chapter is based partly on Palladino, 'Between craft and science'; and 'Wizards and devotees'.
2 Archives of the John Innes Institute (Norwich): William Bateson collection: 2429: Board of Agriculture Memorandum A220/1, September 1911, p. 2.
3 Wilkins, *Research and the Land*, p. 20.
4 For a more detailed account of this formative phase in the history of genetics, see Olby, *The Origins of Mendelism*; and Provine, *The Origins of Theoretical Population Genetics*.
5 University of Reading Archives (Reading): Percival Papers: E. S. Beaven to J. Percival, 20 June 1922.
6 Beaven, *Barley*, p. 5.
7 *Ibid.*, p. 246.
8 Bell, 'The *Journal of Agricultural Science*, 1905–1980', p. 2.
9 'Relation of science to Sunday school teaching', mss in the author's possession, n.d. (ca. 1884), p. 26.
10 Biffen, 'Modern wheats', p. 17.
11 Archives of the Rothamsted Experiment Station (Harpenden): Papers and correspondence of Sir John Lawes and Sir Henry Gilbert, 1897–1901: Vol. 13: E. S. Beaven to Gilbert, 24 October 1898.
12 Beaven, *Barley*, p. ix.
13 *Ibid.*, p. xi.
14 *Ibid.*, p. 249.
15 Archives of the Warminster Historical Society (Warminster): Anonymous, 'Dr Beaven's work honoured', *The Warminster Journal*, 1 July 1930.
16 British Library Manuscripts Collection (London): Alfred R. Wallace collection: 46438: E. S. Beaven to Wallace, 4 February 1911.
17 Hall, *A Pilgrimage of British Farming, 1910–1912*, p. i.
18 E. J. Butler in the discussion following Dampier, 'Agricultural research and the work of the Agricultural Research Council', p. 72.
19 Beaven, *Barley*, p. xi.
20 PRO: MAF 33/177, University of Cambridge, 11 May 1936.
21 Anonymous, 'E. S. Beaven', p. 7.

4

Genetic practices and the end of the subject

> No one would deny that the mendelian laws are as applicable to the human as to the tall and short peas which Mendel used in his original experiments. Yet Mendel could not have formulated the laws if he had not studied an obvious character in a quickly growing plant. Similarly, in the study of breast cancer the use of inbred mice is an invaluable aid to the elucidation of the problem in man.

Of mice and men

It seems hard to remember, but just ten years ago reading a history of genetics and the political economy of agricultural production was a recipe for boredom. I could not get anyone interested. In fact, when I presented a paper at a conference of economic historians, Sally Horrocks, then a historian interested in the equally invisible food processing industry, noted how I managed to awaken some interest only after I mentioned how much Unilever had paid for the Plant Breeding Institute.[1] Perhaps something significant was afoot and we should have been paying closer attention to what was happening in the world of agriculture, but funding bodies simply were uninterested. The secrets about the manufacture of the food we eat, which were revealed in the midst of the public furore over mad cows, genetically modified foods and the foot-and-mouth epidemic, were, and by and large still are, secrets. In the last two chapters, I have sought to unravel one of the reasons for such secrecy: food production is such a mundane and immemorial activity that it is literally invisible. Medicine, on the other hand, was as interesting as it has always been, and especially so as the decoding of the human genome and its biomedical applications appeared to threaten the existence of the human 'as such'. I then turned my attention to medicine, making the same assumptions that Sir Walter Morley Fletcher had made in the 1930s, namely that medicine was no different to agriculture.

I began to engage with the history of British medicine in the twentieth century, motivated largely by the work of my intellectual sparring partners Roger Cooter and Steve Sturdy.[2] Their emphasis, in their many studies of medicine and the making of the British state, on the conflict between conservative clinicians and modernising medical researchers over the relationship between the medicine of the clinic and the medicine of the laboratory struck a resonant note. I was particularly intrigued by a polemical exchange between Lord Moynihan, the President of the Royal College of Surgeons of London, and Sir Frederick Gowland Hopkins, the President of the Royal Society of London and Fletcher's former academic mentor. In 1930, Moynihan rebuked those who believed that,

> Though the method of Hippocrates, of observation . . . has rendered valuable service to Medicine in the past, the future advance of Medicine rests with laboratory workers.[3]

He then evoked the historical achievements of William Jenner to argue that the human body was far too complicated to be as easily analysed as 'laboratory workers' seemed to suppose. These 'laboratory workers' were in fact,

> Lagging too far behind, concerned too much with the laboratory and with mice, too little with hospital wards and with men. We, the

clinicians, have gained gigantic intellectual victories . . . with barely recognisable assistance from the laboratory.[4]

The continued progress of medicine, in other words, rested with clinicians and their historically tested practices. Gowland Hopkins thought otherwise. Echoing the rhetoric of *Science and the Nation*, he responded the following year by arguing that,

> [Moynihan's] complaint that laboratory activities have become too remote from actuality is not justified. He forgets that in the progress of science the emergence of some one fact which has immediate practical application may inevitably need the previous acquisition of knowledge which itself may have no immediate practical value. In solving a practical problem, let us say the control of cancer, the frontal attack is seldom successful.[5]

Then, just like Moynihan, he evoked Jenner's achievements, but this time to praise the laboratory. He claimed that,

> Knowledge of nature is gained both by observation and experiment, but progress by means of the former, and much older, method is almost always slow, while by the latter, and much newer, method it may sometimes be very fast. The essence of an experiment, of course, is that you narrow an inquiry to a single issue, that in studying phenomena you vary only one contributory factor at a time. You study the phenomenon in the presence of that factor and in its absence; and you get an unequivocal answer as to the influence and importance of that factor . . . Forgive me if I am putting the obvious before you; it is a point which sometimes needs emphasising. I would emphasise it by referring to a chapter in the history of medicine. In the middle of the last century this country possessed physicians who were exceptionally fine observers. Among them was William Jenner. Now Jenner had been striving to make it clear to his English colleagues that typhus and typhoid fevers were distinct pathological entities . . . It cost Jenner many years of work to convince his contemporaries by the evidence he produced. Jenner felt the ambiguity of nature's chance remarks but saw apparently no remedy save the

laborious recording of her spontaneous utterings and a painful effort at interpretation. Nature, however, can be taken into the witness box and directly questioned . . . This is the method of experiment . . . [and] only the method of experiment can yield rapid progress.[6]

I did note how this exchange raised important questions about the interested nature of historical remembrance, especially as I came across Ulrich Wiesing's work on Robert Koch's own understanding of the importance of the history of medicine in the making of the physician. What struck me most vividly, however, was the close resemblance of this exchange and the arguments between Sir Rowland Biffen, on the one hand, and John Percival and Edwin Sloper Beaven, on the other hand. Here again was a practitioner calling into question the meaningfulness of the artificial conditions necessary for an experimental analysis of practical problems. Yet, figures such as Moynihan and Gowland Hopkins were already preoccupying more established historians of medicine such as Harmke Kamminga, Mark Weatherall, and Chris Lawrence. As John Harley Warner has noted, exchanges such as that between Moynihan and Gowland occupy a particularly important place in the work of historians of medicine in the nineteenth and twentieth century. They have sought to explain their genesis by pointing out firstly that clinicians such as Moynihan, and laboratory workers such as Gowland Hopkins, studied very different objects. The former focused on human pathology as an end in itself. The latter focused on the normal and abnormal conditions of animals in the laboratory, as means to an end. Secondly, these different forms of study generated and sustained equally different forms of knowledge. On the one hand they generated and sustained the clinician's particular, personal understanding of a patient's condition, grounded in the case study, against the laboratory worker's more abstract understanding of pathological conditions, grounded in the articulation of a 'model' applicable to humans and animals alike. Thirdly, this knowledge evolved in very different institutional settings, namely hospitals as opposed to universities or specialised research institutes. Lastly, in England, the clinicians and laboratory workers elaborating these two forms of work and knowledge were associated with the Royal Colleges of Surgeons and Physicians, on the one hand, and organisations such as the Medical Research Council and the Royal Society on the other hand. Moreover, in linking these contrasts causally and hierarchically from the institutional context to the epistemic content of polemical exchanges such as that between Moynihan and Gowland Hopkins, some of these historians have then contributed to the extension of the social constructionist programme to the history of medicine. This was, in other words, familiar intellectual territory, and I then busied myself searching for more obscure figures not yet considered by these historians. I readily found one in another of Fletcher's objects of vitriol, Percy Lockhart Mummery, senior surgeon at St Mark's Hospital and secretary of the British Empire Cancer Campaign.

Although he was very interested in the familial, if not genetic, nature of cancer, Lockhart Mummery condemned the use of the inbred mouse in the experimental study of cancer. Thus, in 1940, he argued that,

> Experimental results cannot be applied too closely to the problem in man, because the conditions of mating necessary to demonstrate [genetic influence] in mice never obtain in any civilised community of mankind.[7]

Although the public record suggests that Lockhart Mummery's earlier, more general criticisms of 'laboratory workers' went unnoticed, this time Georgiana Bonser, in the Department of Experimental Pathology and Cancer Research at the University of Leeds, rebutted that inbred mice were not nearly as problematic as Lockhart Mummery claimed. She wrote to the *British Medical Journal*, arguing more specifically that,

> No one would deny that the mendelian laws are as applicable to the human as to the tall and short peas which Mendel used in his original experiments. Yet Mendel could not have formulated the laws if he had not studied an obvious character in a quickly growing plant. Similarly, in the study of breast cancer the use of inbred mice is an invaluable aid to the elucidation of the problem in man.[8]

Instantiating the assimilating logic of genetics, humans, peas and mice were, for Bonser, perfectly comparable forms of life. More importantly, the encounter between Lockhart Mummery and Bonser then neatly encapsulated the disagreement between Moynihan and Gowland Hopkins, though on the more familiar terrain of genetics. Replicating Bonser's epistemic elision and Fletcher's institutional elision of all difference between agriculture and medicine, I had thus found my niche in the history of medicine. At the same time, however, my work on Biffen, Percival and Beaven had raised too many questions about the hermeneutic practices of social constructionism, especially its location of agency in social institutions, just to settle for another case study in the social construction of medical knowledge.

Thinking about the organisation of cancer research

During the first decades of the twentieth century, cancer became the new medical problem for the new century. Discussions within the medical profession about the truth of its increase, about its greater diffusion among urban rather than rural dwellers, and about its relative rarity among the 'lesser civilised' people of Africa, clearly reflected a vision of cancer as the disease of modern society. Only modern modes of medical intervention could possibly remedy such an important disease.

As Joan Austoker and David Cantor have noted, hospitals especially dedicated to those suffering from cancer had been established throughout the nineteenth century, and surgical removal of cancerous growths was the main, if often ineffectual, treatment. Cancer was a deadly disease. At the turn of the century, however, the emergence of radiotherapy provided a new and exciting alternative to surgery. Moreover, increasingly assertive bacteriologists and physiologists argued that more than just the surgeons' traditional attention to morbid anatomy was required to hope for greater therapeutic success against cancer. Cancer then became a hotly contested domain, especially as leading figures in the medical profession mounted a campaign to organise a Cancer Research Fund, which was so successful that it obtained royal support and thus resulted in the establishment of the Imperial Cancer Research Fund. Bacteriologists and physiologists soon dominated the Imperial Cancer Research Fund, determining its orientation toward the experimental study of cancer, and the more strictly clinical problems of radiotherapy and the improvement of surgical practices were left to the surgeons in the major cancer hospitals. This compromise, however, was destabilised by the introduction of radium as a new and more easily managed source of radiation. The high cost and government control over relatively great amounts of radium provided the newly established Medical Research Council with the leverage to become one of the major actors in the political economy of health care. The Medical Research Council, under the skilful direction of the yet unknighted Walter Morley Fletcher, quickly proceeded to wrestle away from surgeons the control they enjoyed over radiological practices, to establish these as autonomous investigations of the physiology of irradiated organic tissue. As Cantor has put it, the field of 'radiobiology' was born. In 1923, the increasingly marginalised Royal Colleges of Physicians and Surgeons, which initially had been very sympathetic toward the Imperial Cancer Research Fund, reacted to this situation by supporting the establishment of a second organisation dedicated to cancer, the British Empire Cancer Campaign. Unlike the Imperial Cancer Research Fund, the British Empire Cancer Campaign planned to support work in clinical departments. Fletcher quickly and effectively toned down this greater clinical orientation by rallying the Medical Research Council and the Royal Society to demand that the Board of Trade should refuse the British Empire Cancer Campaign the charitable status it needed to raise funds on a significant scale. Recognising the power of their opponents, Percy Lockhart Mummery and other representatives of the British Empire Cancer Campaign agreed to allow the Medical Research Council and the Royal Society substantial control over the development of research within their organisation. Yet, surgeons such as Lockhart Mummery still remained very powerful figures in the political economy of health care. However financially straitened their hospitals may have been during the years after the First World War, the reach of their network of personal connections was far greater than any institution dedicated to experimental research, and the British Empire Cancer Campaign soon became much wealthier than the Imperial Cancer Research Fund. Not surprisingly, therefore, the arguments over the relative merits of medicine of the laboratory and medicine of the clinic continued long after these initial conflicts. In fact, if the exchanges between Lord Moynihan and Sir Frederick Gowland Hopkins are anything by which to go, these arguments became increasingly acrimonious.

However compelling, Cantor's and Austoker's institutional narratives seemed to me to ignore, if they did not actually dismiss, any active role for clinicians' and laboratory workers' distinctive material practices. These practices simply figured as either hindering or abetting institutional positioning, and their evolution was a consequence of conceptually prior shifts in the institutional balance of power. The disruptive role of radium itself, for example, passed without comment. Reducing this perspective almost *ad absurdum*, it seemed to imply that what people do is completely determined by their location within particular networks of social relationships, rather than by their relationship to the material world. I was then attracted to Jean-Paul Gaudillière's efforts to explain how new tools of inquiry, such as the inbred mouse that so incensed Lockhart Mummery, were created and acquired their current, central role in research on the causes of cancer.[9]

At a workshop to which I was invited, as a new member of the confraternity of historians of medicine, Jean-Paul argued that geneticists' efforts to find a niche in the growing institutional organisation of bio-medical research were mediated by the radical transformation of the inbred mouse. The inbred mouse had originally been introduced as a 'model' to study the relationship between heredity and cancer, but soon became a useful 'reagent' to test chemical therapeutics. This new function and concomitant circulation of the mouse outside the laboratory then allowed its creators to challenge clinicians' control over research into the causes and treatment of cancer. This approach seemed to me to offer a more catholic understanding of people's place in the world. Yet, it also seemed to me that, among these people, Jean-Paul was not treating clinicians and the laboratory workers who sought to displace them in quite the same manner. I was particularly struck by the manner in which he accounted for the unsuccessful criticism of the inbred mouse articulated by Maud Slye, in the Medical School of the University of Chicago. In 1915, Slye argued that cancer was a genetic disease whose inheritance accorded with the mendelian theory of heredity. Clarence Little, of the Cancer Commission at Harvard University, criticised her for having used mixed rather than inbred mice. Slye responded by arguing, as Lockhart Mummery would do in 1940, that inbred mice were inappropriate because they and their artificially induced cancers were radically different from the human situation, where populations were genetically mixed and cancers were spontaneous. Jean-Paul explained

Slye's failure to destabilise the use of inbred mice by appealing to Slye's support for outmoded eugenic ideas, rather than to those details of material practice that were meant to explain Little's institutional successes. Long intrigued by the aesthetics of symmetry, though not in the sense demanded by constructionism, social and otherwise, I wished to avoid such an asymmetric analysis. Ironically, I failed to take note of my earliest work on the chemistry of life, in which I had argued that the aesthetics of symmetry is rooted in the modern suspicion of all that is historically contingent or dependent on the historically contingent.[10] I thus proposed to treat laboratory workers' and the clinicians' distinctive material practices as equally involved in establishing the contemporary equation of mice and humans. Given their disagreement over the use of inbred mice, it seemed to me that Lockhart Mummery and Georgiana Bonser offered a unique opportunity to do so.

Mice and experimental studies of cancer

In 1925, the future Lord Moynihan, then a senior surgeon in the Leeds Royal Infirmary, was concerned to establish a provincial counterweight to the political hegemony of metropolitan surgeons such as Percy Lockhart Mummery. He then spearheaded a local campaign to raise funds for research into the causes of cancer, which resulted in the formation of the largely autonomous Yorkshire Council of the British Empire Cancer Campaign and a Department of Experimental Pathology and Cancer Research in the University of Leeds. Richard Passey, a pathologist at Guy's Hospital who specialised in problems of chemical carcinogenesis and was one of the first researchers to artificially induce cancer in laboratory mice, was chosen to head the new department. Just as he took up his new post, Passey argued that,

> The solution to the problem . . . [of cancer is] hardly likely to come from the clinical side . . . [A]t present the yield of facts from that side . . . [is] getting poorer year by year.[11]

Passey insisted therefore that all prospective members of his new department should be affiliated with the University of Leeds rather than its Medical School or the Leeds Royal Infirmary. Strikingly, the Yorkshire Council of the British Empire Cancer Campaign accepted Passey's demands without any question or comment.

Georgiana Bonser was one of Passey's first appointments to the Department of Experimental Pathology and Cancer Research, though, as a woman, all that she could expect was a temporary fellowship. She was a very promising recruit. She had graduated from the Medical School in the University of Manchester, writing a dissertation on morbid anatomy. After toying with the idea of becoming a surgeon, and being discouraged from doing so, she decided to further her studies in the prestigious Institut Pasteur in Paris. I can only offer it as a matter of speculation, based on Richard Burian, Jean Gayon, and Doris Zallen's study of French genetics, that it was here that Bonser was exposed to the importance of genetic considerations in the management of anatomical studies. This may then explain why, upon taking up her post in the University of Leeds, Bonser immediately recommended that Passey should use inbred mice in his studies of chemical carcinogenesis. Presumably, she hoped to simplify thereby the difficulties Passey was facing in replicating in mice the suspected carcinogenic effects of the soot usually coating workers in the blast furnaces of Yorkshire from head to toe. Such experiments were perhaps unproductive because the mice that Passey used were not in fact susceptible to skin cancer, and thus no amount of tar would produce any meaningful results. Mice that were either invariably susceptible or completely immune to cancer, however, had to be obtained from the Jackson Laboratory, where Clarence Little was by then overseeing the production and marketing of mice specifically designed to study the process of carcinogenesis. Passey had reservations about both the cost and utility of these special mice, and thus he encouraged Bonser to begin breeding her own strains to advance her own preferred studies of the then hotly debated relationship between estrogenic hormones and the process of carcinogenesis.

Bonser clearly enjoyed the professional support of both Passey and Matthew Stewart, the professor of pathology at the University and the Medical Research Council's representative on the Scientific Advisory Committee to the British Empire Cancer Campaign. This, however, was not translated into a satisfactory institutional position within the Department of Experimental Pathology and Cancer Research. Even though she was financially independent, she then took charge of the 'quite undeveloped' pathological laboratory in the less than prestigious Leeds Public Dispensary.[12] Bonser's close friendship with Clara Stewart, Matthew Stewart's wife, one of the leading female medical practitioners in Leeds, and one of the founders of the Medical Women's Federation, perhaps strengthened her resolve to fulfil her professional ambitions by seeking this first foothold in the world of 'clinical medicine'.[13] More importantly, if Bonser's recollections are anything to go by, the concomitant exposure to the complexities and idiosyncrasies of human pathology reminded her of the importance of establishing the comparability of experimental results obtained with her mice and findings in humans. This problem became particularly acute in 1937, when William Cramer and Eric Horning, at the Imperial Cancer Research Fund, criticised the use of inbred mice to study the causes of cancer. Reflecting on William Gye's controversial work on viral carcinogenesis and on their own comparisons of the carcinogenic effects of estrogenic hormones in inbred and genetically mixed mice, Cramer and Horning argued that cancer in inbred mice,

> Appears as a general disease with local manifestations, a disease which cannot be cured by operative removal of the tumour. In animals of . . . mixed strains, cancer appears as a local disease, curable by operation, as in the human subject. It has been stated repeatedly in recent writings that the experimental investigation of cancer should be carried out exclusively in genetically pure strains. From certain points of view this has great advantages. But it also has the disadvantage that it distorts the clinical picture of the disease as it appears in man. . . . It may . . . be questioned whether such highly inbred animals can be accepted as representing normal organisms.[14]

Bonser responded to Cramer and Horning's claim by undertaking a series of experiments, which suggested that the surgical removal of cancerous breasts was equally effective in inbred and genetically mixed mice, and that inbreeding did not in fact qualitatively alter the situation. Cramer and Horning's questions about the clinical relevance of experimental investigations in inbred mice were unfounded. There thus was no reason not to exploit the practical 'advantages' presented by the use of inbred mice, these being chiefly 'their short-life span and . . . ease with which a large number of tumour-bearing animals can be obtained'.[15] In other words, Bonser espoused the values of precision, scale and efficiency discussed at great length by Steve Sturdy and Roger Cooter in their study of science, scientific management and the transformation of British medicine during the first half of the twentieth century. Significantly, these were exactly the same values that John Percival and Edwin Sloper Beaven called into question in their critiques of the mendelian theory of heredity.

These values of precision, scale and efficiency were particularly evident from 1940 onward, as Bonser began to exploit the 'advantages' of inbred mice in a novel context. The Medical Research Council, thanks to Matthew Stewart's intervention, called on Bonser to aid the evaluation of a hormonal treatment for cancer that had once been recommended, on clinical grounds, by Moynihan and other members of the Royal College of Surgeons. The Medical Research Council was sceptical about clinical modes of evaluation, viewing them as less than 'scientific'. As Bonser embarked on this new line of work, she stressed the importance of working with inbred mice, to assess more exactly the optimal dosages that would result in physiologically meaningful tests of this treatment. At the very same time, and perhaps more importantly, such exactitude was also evident as she began to engage with Wilhelm Hueper's highly controversial claims about the link between industrial pollution and cancer of the bladder. Bonser lent greater weight to Hueper's findings as she confirmed them by using genetically 'pure' mice and chemically 'pure' hydrocarbons. This meant that, when ICI, the British counterpart to Hueper's American nemesis, DuPont de Nemours and Co., became increasingly concerned about the proliferation of cancer of the bladder among its own workers, which raised the prospect of lawsuits for workers' compensation, it turned to Bonser for assistance. Ironically, by then funding a research chemist who would collaborate with Bonser, ICI provided her assistant with the kind of institutional support she herself still did not enjoy. The most significant outcome of this collaboration was, however, the establishment of Bonser's novel strain of inbred mice, 'IF', which reacted very rapidly to local and distant exposure to chemical carcinogens, as a very effective diagnostic tool for exposure to environmental carcinogens.

This transformation of Bonser's mice and their increasingly wider circulation outside the world of strictly experimental studies of carcinogenesis, into the world of occupational health, had important institutional repercussions. In 1948, Bonser finally obtained a permanent teaching post in the Department of Experimental Pathology and Cancer Research and was appointed as consultant in morbid anatomy at St James' Hospital in Leeds. Four years later, Hueper repaid Bonser's earlier support by inviting her to speak on chemical carcinogenesis at both the National Cancer Institute and the prestigious Gordon Research Conferences. Bonser's rapidly growing prestige within the medical profession eventually resulted in her election to the Royal College of Physicians in 1954 and, following in Clara Stewart's footsteps, the presidency of the Medical Women's Federation in 1957.

Significantly, during these years, Bonser began to change her preferred experimental organism from inbred mice to genetically mixed rabbits and dogs because the physiology of occupational cancer was studied more easily in these animals. The results of this work were less than satisfactory. The challenge grew even greater as she insisted ever more explicitly that the exposure of workers to industrial chemicals should be viewed as a unique opportunity for the 'experimental' study carcinogenesis in the 'human subject'.[16] Since she did not know the genetic constitution of these new 'objects' of inquiry, Bonser could never be as sure that any cancers were indeed due to these industrial chemicals as she was with her genetically neutral 'IF' line. It is then not surprising that Bonser became much more interested than she had ever been earlier in Mary Wainmann's statistical studies of breast cancer in women around Leeds. The statistical approach to establishing a more tightly controlled assessment carcinogenesis in the 'human subject' was, however, still unproductive by Bonser's exacting standards. Such unproductivity, moreover, could destabilise her carefully nurtured advancement of the inbred mouse as the ideal object for the study of carcinogenesis, as her collaboration with Wainmann revealed no inherited susceptibility to breast cancer comparable to that found in mice. Bonser then redoubled her efforts to establish the equivalence of carcinogenesis in mice, rabbits and dogs, on the one hand, and the 'human subject', on the other hand. These efforts continued until well after Bonser's retirement in 1963.

Everyone eventually forgot Bonser, the medical researcher and champion of medical women, except perhaps for a historian, who found himself digging through archives and interviewing anyone who might have known Bonser. Bonser's mice and the unpromisingly entitled paper in which she described their properties, 'The carcinogenic properties of 2-amino-1-naphthol hydrochloride and its parent amine 2-naphthylamine', are far from forgotten, however. Bonser's 'IF' line still plays an important role in the world of what is now known as 'environmental health'.[17]

In sum, Bonser's development of inbred mice as a tool for cancer research suggests that the boundaries between the institutional settings for clinical and experimental medicine are not as sharp as might be imagined by a social constructionist. If Cramer's and Bonser's work reflected the orientation of their respective institutional patrons, the 'experimentally' oriented Imperial Cancer Research Fund seemed less supportive of the use of inbred mice lines for experimental studies of cancer research than was the more 'clinically' orientated British Empire Cancer Campaign. If Cramer and Bonser were not so representative, this should raise questions about the arbitrariness of the alignments of material practices and institutions used to constitute the social constructionist division between clinicians and experimental scientist. In other words, the material practices of cancer research are indeed as disruptive as Jean-Paul Gaudillière suggested in his own study of developments in the United States. More importantly, however, Bonser's transformation of the mouse into a useful tool for clinical work does not in fact help us to understand why clinicians ever accepted her equation of mice and the human patient. As Bonser herself admitted in recollections of her career, maintaining the equation was 'difficult and exacting', to say the least.[18] As I then argued with Jean-Paul, we needed to evaluate the development of clinical medicine on its own terms, attentive, however, to Bonser's own preoccupation with establishing a different understanding of the 'human subject'.

The modernisation of surgery

During the controversies surrounding the establishment of the British Empire Cancer Campaign, Walter Morley Fletcher and Sir Frederick Gowland Hopkins, characterised Percy Lockhart Mummery as the quintessential representative of the grasping and scientifically illiterate clinicians from Harley Street.

Lockhart Mummery clearly was one of the major figures on Harley Street. It was said that there was no one among the aristocracy of London whom he had not treated, which is quite believable since one of his areas of specialisation was the humble and common haemorrhoid. Equally clearly, his approach to medical research was motivated by his desire to improve his reputation as a superior surgeon. For example, he obtained funding from the British Empire Cancer Campaign and the Medical Research Council to catalogue and standardise records of the tissues studied in the pathological laboratory at St Mark's Hospital. Cuthbert Dukes, the pathologist who was entrusted with the task and then developed the 'ABC' classification of cancers of the colon, once recalled that Lockhart Mummery's intention was to resolve argument between himself and Ernest Miles over the merits of their respective surgical interventions to remove intestinal and rectal cancers. Whatever the merits of his particular mode of intervention may have been, however, Lockhart Mummery's reputation depended much more crucially on detecting the cancers as early as possible. This was facilitated by his improvement of the electric sigmoidoscope, a rectal probe, which he celebrated by writing that,

> The human mind is so constituted that we always experience a certain pleasure when we have succeeded in obtaining access to some spot hitherto difficult or impossible to explore. And in medicine and surgery the perfection of a method by which we are enabled accurately to explore hitherto inaccessible portions of the human body always arouses our interest and helps to perfect our art.[19]

Yet, the reach of the sigmoidoscope was quite limited. Another way to inspect the colorectal tract for unusual growths was to predict their development long before they were betrayed by the symptoms of their cancerous transition, diarrhoea and anal bleeding. Lockhart Mummery knew that Harrison Cripps, a surgeon at St Bartholomew's Hospital, had argued in 1882 that 'polyposis intestini', a condition that eventually led to cancer of the colon, was an inherited condition. Knowledge of familial patterns might then aid early detection. Cripps' claim, however, rested on the observation of just two cases. Lockhart Mummery was then very impressed by Maud Slye's experimental demonstration that cancer was an inherited condition. He thus wrote favourably about her work on the pages of the *Lancet*, notwithstanding its much more critical reception among geneticists and clinicians interested in the inheritance of human pathologies. Perhaps there was much truth to Fletcher and Gowland Hopkins' suspicions about Lockhart Mummery: his interest in the scientific investigation of the causes of cancer was motivated, and undermined, by economic self-interest.

On the other hand, Lockhart Mummery's response to Slye's work was, in fact, far from uncritical. He also argued that one needed to pay attention to the mechanisms that brought about the cancers. Detecting patterns of inheritance was, by itself, meaningless. With further funding from the British Empire Cancer Campaign, Lockhart Mummery then charged Dukes with the collection of family histories of patients suffering from 'polyposis intestini'. By 1931, Lockhart Mummery advanced a more complex theory of inherited susceptibility to local somatic mutations to explain the observed histories of the cancer.

For all his growing interest in cancer and heredity, however, Lockhart Mummery rejected any claims about the nature of cancer that were grounded in studies of cancers in inbred mice. In 1933, for example, Lockhart Mummery became embroiled in a polemical exchange between his friend William Sampson Handley, senior surgeon at the Middlesex Hospital, and James Murray, the director of the Imperial Cancer Research Fund. During a meeting at the Royal Society, on 'experimental production of malignant tumours', Handley echoed Lord Moynihan's criticism of 'laboratory workers' by calling into question the clinical significance of theories about the causation of cancer that were grounded in experimental studies of animals, especially mice. Lockhart Mummery wrote in the *Lancet* that Handley had drawn attention to

> A cause of very real danger to the best interests of cancer research in the tendency there is for animal experimentalists to get out of touch with the surgeons and pathologist, who are dealing with the disease in human beings.[20]

Lockhart Mummery would seem, therefore, to fit more closely the sociological picture of élite clinicians seeking to preserve their professional domain from the incursions of experimental scientists like Fletcher, Gowland Hopkins, and Georgiana Bonser. In fact, the preface to Lockhart Mummery's essay on the future of humanity, *After Us, or the World as it Might Be*, suggests that he was a very close friend of one of the chief bastions of the conservative clinical establishment during the inter-war years, Lord Horder.

At the same time, however, it is also obvious that Lockhart Mummery was far from a conservative opponent of the reorganisation of medicine who refused any truck with modern rationality. In *The Origin of Cancer*, his fullest exposition of his theory about the relationship between cancer and heredity, Lockhart Mummery wrote that,

> Medicine is only just beginning to become a science, and we are only just commencing to understand how the normal human body works, and the manner in which those variations from the normal, which we are accustomed to call disease, take place. While most other sciences have been drawn in to assist medical knowledge, and more particularly the treatment of disease, the true science of medicine, that is to say, the study of the real causes of diseased states of the human body, is still in a very inexact and chaotic condition. As so well pointed out by Wilfred Trotter in his address 'De Mininimis': up to the present time medicine has almost wholly avoided the burden of measurement, but a time will come, and is now indeed coming, when an exact and exhaustive numerical exploration of the facts of disease will have to be undertaken.[21]

He then proceeded to prepare the ground for a discussion of his theory of somatic mutation by discussing advances in cell biology and genetics, concluding that,

> At last medical science is getting down to the fundamental facts of life and growth, and this will lead inevitably to that knowledge of the primary causes of disease and abnormality which is so necessary if we are able to satisfactorily control disease in human beings.[22]

In practice, this programme to transform the organisation of medicine through science was to be achieved by encouraging the Dukes to study both normal and abnormal conditions. This might, sometimes, even imply studying mice, something that Lockhart Mummery does not seem to have ever opposed in the principled way suggested by the critical comments cited earlier. In his first, published discussion of the influence of heredity on cancer, Lockhart Mummery referred approvingly to Slye's work, even though it was based in studies of over five thousand mice. His own theory about the relationship between cancer and heredity rested on experimental studies of mice, rabbits, and fowl. More paradoxically still, at a conference organised by the British Empire Cancer Campaign, he proposed during a session on 'future avenues of research in relation to cancer' that efforts should be made to establish 'factories' to produce mice with 'spontaneous and tar cancers'. Moreover, specially trained technicians should run these 'factories', so to save work for 'medically or scientifically trained laboratory workers'.[23]

Much more importantly, the collection of family histories on which Lockhart Mummery's theory about the relationship between cancer and heredity rested, relied not just on formal referrals to St Mark's Hospital, but also on a network of consultants across the nation, who had passed through the Hospital as registrars. As they learned from Lockhart Mummery a new mode of specialist practice in which patients were no longer treated as the private property of individual consultants, these patients became exemplars of a shared, larger diseased group whose investigation might reveal something about the mechanics of carcinogenesis more generally. That is to say, human themselves had become 'models', and in this sense quite comparable to mice. Admittedly, it could be argued that, since there already was a tradition of collaborative investigation of disease stretching back to the late nineteenth century, this transformation was not so novel. On the other hand, the sharing of records was still resisted in St Mark's Hospital, and elsewhere, well into the 1950s. Furthermore, those consultants who did submit their records were no longer active participants in the investigation, a role that would have preserved at least the semblance of personal connection with their patients. The labour of discovering the genetic causes of human disease was being 'divided', if not yet 'de-skilled'.

In sum, Lockhart Mummery's work at St Mark's Hospital amounted to the realisation of Lord Moynihan's vision of a science of the clinic that was fully comparable, if not superior, to the medicine of the laboratory. Moynihan had advocated as a first step toward the realisation of this vision that teachers of medicine should 'inculcate the virtues and to instruct in the methods of observation as the primary faculty required in students'.[24] This would allow medical men of the future to overcome the

> Impossibility of translating in every case, or under all conditions, the animal experiment into terms strictly applicable to man and establish the conditions for *hominal* . . . experiment and research.[25]

Lockhart Mummery had already begun to realise these conditions to improve the efficiency of his surgical interventions. If Bonser transgressed the boundaries between humans, mice and peas, Lockhart Mummery transgressed those between the individuated human being, *bios*, and *zoē*, the human being in its species existence.

Do humans make a difference?

While I was still working on the history of agricultural genetics, Steve Sturdy was busy writing a critical review of Bruno Latour's *Pasteurization of France.* He claimed that, however compelling philosophically, Latour's argument for a symmetric approach to the natural and social worlds, and the rejection of causal explanation, was politically disempowering. Steve eventually wrote with aplomb that such a position spoke of 'a cenobitic impulse that leads [Latour] to favour the security of philosophical contemplation over the uncertain and potentially compromised business of producing knowledge about the world'.[26] As Steve, Roger Cooter, Mary Fissell, Jonathan Harwood, John Pickstone, and I argued endlessly about the merits of Latour's ideas, I began to find Latour's and Steven Woolgar's criticisms of social constructionism as drawing an arbitrary and unwarranted distinction between the natural and social sciences appealing. If, on the one hand, I wanted to be contrary, simply to assert my own agency against the hegemony of social constructionism, on the other hand, it seemed to me that social constructionism was as reductive as the medical sciences it sought to criticise. Moreover, contrary to the implicit thrust of Steve's argument, it provided no political justification for preferring one form of reductionism to the other. The overt political commitments that had originally motivated social constructionism, especially as it turned its attention to medicine, as in Peter Wright and Andrew Treacher's *The Problem of Medical Knowledge*, were a distant memory. Even the combative Roger, who once contributed signally to *The Problem of Medical Knowledge*, was preparing to throw up his hands in despair before the contemporary transformation of medicine, as it blurred the secure boundaries between the natural and the social, by writing that:

> Moral legitimacy is up for grabs. 'Dolly', a cloned lamb with a tart's name, serves perhaps as a meaningful symbol of the uncertain, ambivalent moral precipice on which we now stand.[27]

I thus saw no reason not to prefer Latour's more philosophically compelling argument.

This said, my response to Woolgar and Latour's argument for a symmetrical treatment of the social and the material worlds was one of 'misrecognition', the productive recognition of self in an image that is in fact not identical with the self. Although I persisted in centring my work on the human subject, in a manner that was completely contrary to Woolgar and Latour, their argument seemed to offer me an opportunity to do something different from my debating partners. It seemed to offer me an opportunity to acknowledge the agency of the material practices by which we differentiate various human activities.

Following Latour's Archimedean aphorism 'give me a laboratory and I will raise the world', a number of historians of the bio-medical sciences were beginning to view the practices of the laboratory as the critical locus of study.[28] As the many contributors to Adele Clarke and Joan Fujimura's *The Right Tools for the Job* noted, it was here that the boundaries and relationships between the natural and the social orders were forged, and here that the elision of difference between mice and humans was established. It seemed to me, however, that, in its effects, this approach was complicit with the triumphal histories of contemporary bio-medical science. I wished to return some symmetry to the argument by focusing on the clinic. This enabled me to extend Chris Lawrence's argument in *Medicine in the Making of Modern Britain* that by the 1920s a modern outlook, grounded in a scientific approach to the enterprise, had become hegemonic in British medicine. What was at stake in the arguments between Lord Moynihan and Sir Frederick Gowland Hopkins, or William Sampson Handley and James Murray, or even within the supposedly clinically orientated British Empire Cancer Campaign, between Percy Lockhart Mummery and Georgiana Bonser, was the locus for the new medicine. It was not a conflict over some essentially different nature of clinical and experimental expertise, which might perhaps have justified the usual labels 'conservative' and 'modern'. For Lockhart Mummery, as for Moynihan and Sampson Handley, the new, scientific medicine, in which humans and mice were not very different objects of investigation, was to be developed in hospitals rather than specialised research institutes or universities.

In sum, the arguments over the relative merits of medicine of the laboratory and medicine of the clinic were arguments on the margins of that fundamental reconfiguration of the 'human subject' to which Bonser spoke. This, however, began to raise questions about the word 'subject', which I had perhaps been deploying far too uncritically.

Notes

1 For an introduction to Sally Horrocks' work, see 'Nutrition science'. This chapter is based partly on Palladino, 'On writing the histor(ies) of modern medicine'.
2 For an introduction to Steve Sturdy's and Roger Cooter's work, see Sturdy and Cooter, 'Science, scientific management, and the transformation of medicine in Britain, 1870–1950'.
3 Moynihan, 'The science of medicine', p. 785.
4 *Ibid.*, p. 784.
5 Gowland Hopkins, 'The clinician and the laboratory worker', pp. 206–7.
6 *Ibid.*, p. 208.
7 Lockhart Mummery, 'Summary', p. 17.
8 Bonser, 'Influence of heredity on breast cancer', p. 456.
9 For an introduction to Jean-Paul Gaudillière's work, see Gaudillière, 'Circulating mice and viruses'.
10 The pivotal place of 'symmetry' within social constructionism is discussed in Bloor, *Knowledge and Social Imagery*, pp. 175–9.
11 Passey, 'Cancer', p. 653.
12 Contemporary Medical Archives Centre (London) (henceforth CMAC): SA/MWF/C102: Georgiana Bonser: Curriculum vitae, n.d. [1970?], p. 3.
13 *Ibid.*
14 Cramer and Horning, 'Adrenal changes associated with oestrin administration and mammary cancer', p. 638.
15 Bonser, 'The value of inbred mice in relation to the general study of mammary cancer', p. 125.
16 Bonser, 'Recent trends in cancer research', p. 23.
17 See http://potency.berkeley.edu/text/CPDBreference.html
18 CMAC: SA/MWF/C102: Georgiana Bonser: Curriculum vitae, n.d. [1970?], p. 3.
19 Lockhart Mummery, 'The diagnosis of tumours in the upper rectum and sigmoid flexure by means of the electric sigmoidoscope', p. 1781.
20 Lockhart Mummery, 'The Royal Society discussion on experimental production of malignant tumours', p. 323.
21 Lockhart Mummery, *The Origin of Cancer*, p. 3.
22 *Ibid.*, p. 8.

23 Archives of the British Empire Cancer Campaign (Oxford): Biennial Informal Conferences: Informal conference, 10–11 December 1930.
24 Moynihan, 'The science of medicine', pp. 779–80.
25 *Ibid.*, pp. 780–1 (emphasis in the original).
26 Sturdy, 'The germs of a new Enlightenment', p. 173.
27 Cooter, 'The ethical body', p. 466.
28 Latour, 'The force and the reason of experiment', p. 76.

5

Genes, genealogies and the return of the subject?

> [Explain to him that] the dominant gene may be of low penetrance and that the hardships and intestinal upsets caused by his time as a prisoner of war in Japanese hands may have caused the appearance of overt disease.

> We have sent our beaters out after some polyposis children who have not been seen for a while or not at all. One of these patients has apparently been caught in your net.

A vanishing act

The two epigraphic excerpts with which this chapter opens capture two very different aspects of the relationship between medical professionals and patients that would have greatly interested Mary Fissell.[1]

Mary was perhaps one of the most daunting intellectual figures around me while I was working on the history of genetics and agriculture. She had the unparalleled ability to make many other historians and me squirm anxiously, by asking why anyone should care about the lost worlds in which we immersed ourselves as historians. 'Why should anyone care?' is the simplest, most facile, but none the less important, question any historian has to confront. In the meantime, Mary was writing *Patients, Power and the Poor*, a book on the patient in the organisation of medical care in early modern Britain. In our many discussions of historiographical problems,

Mary introduced me to the idea that, as medicine evolved into its modern form, the patient 'disappeared'. As medicine became an objective science of the ailing body, the patient gradually ceased to be the formative agent he or she had once been. The political 'subject' that once was, became a scientific 'object'. Mary aimed therefore to write politically by seeking to return 'ownership of themselves' to the poor and other 'ordinary' people who suffered most from the emergence of modern medicine.[2] Many of my colleagues and I appeared to fail signally in this task as we instead focused our attention on scientists and scientific institutions, the agents of objectification. Yet, Mary's arguments about the disappearance of the patient resonated oddly as I began to examine the evolution of the genetic understanding of carcinogenesis proposed by Percy Lockhart Mummery.

About seven years ago, a small report in *The Sunday Times* caught my attention. Its gist is captured by the following few lines:

> British doctors will for the first time use a test to select cancer-free babies next month. The procedure raises the prospect of designer babies . . . Embryos of a woman with a high risk of passing on a form of bowel cancer will be screened and only healthy ones will be re-implanted. The same technique is likely to be used within two years to screen test tube embryos for a predisposition to inherited breast cancer.[3]

The 'woman' in question suffered from a medical condition known as 'familial adenomatous polyposis'. Thanks partly to Lockhart Mummery himself, it is now understood as a rare condition, but it has none the less been the subject of much medical investigation. It has often seemed to provide a 'model' for understanding the relationship between cancer and heredity that does not fall into any facile genetic reductionism. The salient characteristic, which makes the condition so interesting to contemporary medical professionals, is some people's inherited tendency to develop, in teenage years, innumerable polyps throughout their colorectal tract. Eventually, some of these polyps will become malignant and lead to cancer of the colon. The initiation of this complex process is attributed to a fully sequenced mutation of the APC locus on chromosome 5q21. Significantly, the final, often deadly outcome of 'familial adenomatous polyposis' is preventable. Parts of the colon can be surgically removed as they become infested with the polyps, but in the case discussed in *The Sunday Times* this preventive intervention also meant that the woman was no longer able to bear children. Artificial insemination of the woman's ova and the screening of the resultant embryos for the 'FAP' mutation, however, promised her a bright future. She would once again be able to bring children into the world, children who would not fall to the cancer that had already killed her mother and two sisters. Yet, the report in *The Sunday Times* also raised the prospect that this noteworthy medical achievement was the first step to the production of 'designer babies'. This phrase encapsulates a common fear that genetic knowledge and its associated reproductive technologies signal a return to eugenics, a fear which until very recently has inclined the Human Fertilisation and Embryology Authority to prohibit pre-implantation embryo screening anywhere in Britain.

I was often struck by the way in which these evocations of a return to eugenics, now more 'medicalised' than was ever the case in the early twentieth century, often portray patients and their families as passive objects of 'professional' intervention. On the one hand, if these interventions were extinguishing the 'human subject', this was a fitting consummation of the narrative of the patient's disappearance. On the other hand, medical professionals seemed absolutely bewildered by the diversity of patients' responses to the genetic services they were, and are, being offered. In fact, around the time of the report in *The Sunday Times*, the *Times Higher Education Supplement* published Gail Vines' interview with

Theresa Marteau, the director of the Psychology and Genetics Research Group in Guy's and St Thomas's Hospital. In this interview, Marteau discussed her efforts to better understand patients' diverse responses to pre-natal screening, efforts that paid particular attention to the developments around 'familial adenomatous polyposis'. As I began my project on the history of medical genetics in Britain, it seemed to me that such concern contradicted Mary's argument that modern medicine is a monological enterprise, in which medical professionals alone speak with any authority. This scepticism, however, can be questioned in at least two ways, both of which would transform the small report in *The Sunday Times* into what the Michel Foucault of 'Questions of method' would call an 'event'.

During a workshop on medicine and genetics, I happened to discuss the historiographical problem presented by the report in *The Sunday Times* with the medical anthropologist Martin Richards.[4] He suggested that patients' responses and their reasoning could only be as 'interesting' as Marteau seemed to find them if they contradicted a tacit, medically rational presupposition that a genetically defective embryo should be aborted. Furthermore, Marteau may point out that genetic considerations are only one class of the many involved in patients' decisions about reproduction, and that, therefore, all that medical professionals like herself should do is to enable these patients to make a more informed choice. Yet, the presumed freedom of choice may be more apparent than real. It has certainly been facilitated by the ongoing reforms of the National Health Service, which aim to reposition patients as 'informed consumers' in a novel medical marketplace where economic rationality reigns supreme. Thus, Sir Walter Bodmer, a renowned geneticist, leading figure in the British effort to decode the human genome and former director general of the Imperial Cancer Research Fund, has recently noted that patients' choices regarding genetic screening for 'familial adenomatous polyposis' are not made in a vacuum. They must be balanced against their cost to the increasingly insolvent National Health Service. Marteau's attentiveness to patients' responses to pre-natal screening for 'familial adenomatous polyposis' does not then invalidate the narrative of a return to eugenics, this time, conducive to logic of advanced consumer capitalism rather than the logic of an earlier corporatist state.

Following Mary, my job as a more politically engaged historian was to unmask the world of coercive, dehumanising institutional power that lay behind Marteau and her collaborators' engagement with their patients. It is notoriously difficult for historians to reconstruct patients' voices, however. Their exclusion from the archive of medical history has been a necessary precondition for the establishment of the medical profession's contemporary authority. My first task was then to search for the counterpart to Mary's parish records, a hitherto undisclosed archive that had somehow escaped the hegemony of modern medicine. This would allow me to return 'ownership of themselves' to the 'ordinary' people who have, and will have, suffered most from the emergence of genetic technologies such as that described by *The Sunday Times.* Yet, as I left my colleagues at the University of Manchester for Lancaster University, I began to discover that Richards' was not the only way to understand the world now being made by medical genetics. Following Foucault, one might argue that, in acting as informed consumers, the patients studied by Marteau are not the objects of professional control evoked by Richards. They are instead constitutive figures of a new 'discourse', in which patients' genetic endowment is an integral part of how they understand themselves. The increasing prominence of patients' groups in the evolution of research programmes into the genetic bases of disease certainly lends great weight to this perspective. What seems even more compelling, however, is the way in which those patients who decline the offer of a genetic test, seem to do so without ever calling into question the legitimacy of what Evelyn Fox Keller, in *Century of the Gene*, has called 'gene talk'. These paradoxically affirming refusals bear witness to the new age of 'bio-power', an age dominated by the internalisation of the disciplinary knowledge and regulatory practices evoked by the injunction 'Only healthy seed must be sown.'

Admittedly, I was particularly attracted to this Foucauldian approach because, if there was anything to be learned from the encounter between Percy Lockhart Mummery and Georgiana Bonser, it was that there is no position outside the logic of modernity. There is no other archive outside the discourse of the 'gene'. If anything could then be said about disappearance, it was not the 'patient' who is disappearing, but the 'human subject' of the modern imagination. At the same time, however, it seemed to me that I perhaps risked dismissing far too quickly and easily the difficulties that obviously frustrated Lord Moynihan's ambition to establish a 'hominal' science. Sir Frederick Gowland Hopkins, as we have seen, understood the relationship between the laboratory worker and the natural phenomena they sought to explain as perfectly transparent. While William Jenner may have been captive to 'the ambiguity of nature's chance remarks', the experimental approach allowed Gowland Hopkins to bring nature to the 'witness box' and put questions so 'that there is no possibility of ambiguity in the answers'. Moynihan was just as committed to this inquisitorial approach, but he also was dissatisfied with the 'the method of experiment' because 'the solitary question posed by the physiologist is answered 'yea' or 'nay', the answer being conditioned by the power of the chosen animal to reply'. Moynihan, reflecting on his encounter with the human 'animal', worried about the perversities of this particular 'witness', writing that

> The observer is . . . confronted not with phenomena as they really exist, but with impure phenomena in varying degree, adulterated by the mind or affected by the will or temperament of the patient.[5]

It seemed to me that Moynihan spoke to something deeply refractory to those objectivising ambitions that would treat plants and patients as fundamentally indifferent forms of life. I then sought to reconstruct a history of 'familial adenomatous polyposis' with an eye that was more attentive to the interaction between the medical professional and the patient.

I began this alternative, somewhat sceptical enterprise with the Polyposis Registry, in St Mark's Hospital. The then emerging controversy over deCODE Genetics' purchase of the medical records of the entire population of Iceland was a salutary reminder that the contemporary promises of molecular genetics were inseparable from genealogical investigations such as those pursued by the Registry. In this particular case, however, it was a business of a couple of inconspicuous rooms, one of which was taken up by a dozen filing cabinets. Unlike any archive that I had previously encountered, these filing cabinets contained the medical records of innumerable patients, as well as these patients' family histories, some correspondence between medical professionals, patients and these patients' relatives, and a number of publications relating to the work of the Registry. A story began to unfold as I tried to weave these very different fragments together.

Conditions of possibility

As I noted in the last chapter, in 1882, Harrison Cripps claimed that 'polyposis intestini' was a familial condition leading to cancer of the colon. This claim was not examined any further for another forty years, when Percy Lockhart Mummery returned to it in the wake of Maud Slye's work, which supposedly demonstrated that cancer was a genetically determined disease. Focusing on the polyps and their frequent, but not determining association with cancer, Lockhart Mummery argued that a hereditary predisposition did not explain what caused the actual manifestation of cancer. He then began to build a more complex theory by using the genealogical data that he had begun to collect in St Mark's Hospital. From 1932 onward, he argued that families suffering from 'familial polyposis' shared a gene that specified an instability of the somatic genetic material, which then led to excessive cellular proliferation and increased chance of malignant mutations among these anomalous cells. As I also noted in the last chapter, the collection of the records, on which this explanation depended, rested on a network of consultants who had passed through St Mark's Hospital as registrars. As they exchanged the records they collected in their own, later practices, the encounter with persons afflicted by 'familial polyposis' ceased to be the private relationship upon which members of the medical élite built their prestigious practices on Harley Street. They became instead exemplars of a shared, larger diseased group, whose investigation might reveal something about the mechanics of carcinogenesis more generally. Significantly, the social interconnection entailed in Lockhart Mummery's method and explanation also meant that cancer became a social disease. This radical transformation of cancer was reinforced by Lockhart Mummery's more speculative discussions of the origin of cancer. It was a disease of a modern civilisation whose failings could only be redressed by the adoption of 'controlled breeding'.[6] Paradoxically, however, Lockhart Mummery never linked his quite radical eugenic beliefs and the treatment of cancer. In fact, he explicitly dismissed the contemporary proposals for the 'preventative' management of the disease articulated by William Cramer, the physiologist in the Imperial Cancer Research Fund who was interested in the relationship between 'intrinsic' and 'extrinsic' causes of cancer, as well as a critic of the use of inbred mice in the investigation of such a relationship. Lockhart Mummery wrote that,

> The chances of being able to prevent cancer on the lines suggested by Dr Cramer is [sic] not a very hopeful one. There is, however, one point worth noting. Where it is known that certain individuals have possibly inherited a susceptibility to develop cancer in a certain organ, then if such individuals are carefully examined, as regards that organ, at regular intervals, there is an excellent chance of the lesion being detected during the early stage, when it is curable.[7]

Social constructionists such as Roger Cooter and Steve Sturdy would argue that this was the result of Lockhart Mummery's professional interest in maintaining the priority of 'clinical medicine' over the discourse of 'social medicine', which focused on statistical norms and populations, rather than the idiosyncratic conditions of individual patients. Such an explanation, however, raises questions about the reasons for Lockhart Mummery's return to Cripps' claim in the first place. To answer them, we must attend to his patients.

Lockhart Mummery had become a renowned surgeon thanks to his private practice on Harley Street. His reputation as a specialist in the treatment of cancer of the colorectal tract depended on detecting it as early as possible. As I have already pointed out, this was facilitated by his improvement of the electric sigmoidoscope, but the reach of the sigmoidoscope quite limited. Another way, implied by Lockhart Mummery's response to Cramer, might be to somehow predict the manifestation of cancer. Studying patients' healthy relatives to detect hereditary patterns, and thus assess the risks might do this. Yet, the aristocratic patients on whose custom Lockhart Mummery's renown rested would have been unwilling to reveal their family secrets about irregular bowel movements and anal discomfort to someone as inferior to them as a surgeon. This was especially so since these healthy relatives would then be asked to submit to a speculative and degrading sigmoidoscopy, which, as Sir Walter Bodmer politely acknowledges, still is an 'unpleasant procedure'.[8] Establishing the genetic nature of colorectal cancer, which might then serve to justify on more polite, scientific grounds an otherwise rude inquiry about these patients' relatives, called for the surveillance of the relatives of the politically less difficult patients. The patients referred to St Mark's Hospital by humble cottage hospitals in London's East End might have been useful substitutes. Unfortunately, however, these last patients were not always willing to co-operate with Lockhart Mummery, and thus repay the civic notables and patrons of St Mark's Hospital, who sometimes interceded on the patients' behalf to win them admission into the Hospital. Similarly, in 1954, Peter Brasher, one of Lockhart Mummery's colleagues at St Mark's Hospital, recalled how twenty years earlier another colleague had sought repeatedly to study one particular patient's relatives, but

> Co-operation . . . was never freely given, mainly because the father believed that all treatment was meddlesome. They would not communicate with their relatives or give their addresses.[9]

The contemporary correspondence between clinicians and their patients, which was necessary to construct the family histories, does not reveal why the latter found the request so 'meddlesome'. We can imagine, however, that they, like Lockhart Mummery's aristocratic patients, did not wish to submit their relatives to the speculative and degrading sigmoidoscopies they themselves had undergone. Given the thus limited data that Lockhart Mummery could mobilise, his genetic explanation of cancer remained an unrealised technology to visualise the clinical body's interior and thus perfect surgical practice, rather than a platform for 'controlled breeding'. Of course, this also means that there is no necessary connection between genetics and eugenic ambitions. Nevertheless, Lockhart Mummery's practices established what Michel Foucault would have called the 'conditions of possibility' for such a connection between knowledge and power.[10]

Disciplinary power and the amplification of dissonance

In 1956, the renowned geneticist J. B. S. Haldane argued, in a much publicised essay on 'the prospects of eugenics', that,

> It is the duty of a physician or surgeon to tell [anyone carrying the gene for 'polyposis coli'] that about half his or her children will at worst die of cancer, at best be condemned to a life of semi-invalidism . . . [S]uch persons should be taught methods of birth control; perhaps they should be given the opportunity of voluntary sterilization.[11]

The momentous conceptual transformation of the now re-named 'polyposis coli' into an exemplary genetic disorder that was also eugenically significant was facilitated by the reorganisation of medicine under the National Health Service. Those charged with this reorganisation questioned the future of St Mark's Hospital as an 'independent special hospital' because they did not view 'proctology', the science of the colorectal tract, as a legitimate medical specialisation. This prompted the creation of a financially independent Research Department, with Cuthbert Dukes as its first director, to reinforce the notion that St Mark's Hospital was none the less an important centre for medical research. At the same time, the National Health Service was committed, at least in principle, to social medicine. While epidemiology is the most notable discipline associated with this form of medicine, genetics was also very important since its approach to understanding, if not treating, disease was fundamentally social. Dukes' appointment was due largely to his alignment of polyposis, genetics, and social medicine.

Percy Lockhart Mummery had brought Dukes onto the staff at St Mark's Hospital in 1922, to take over a new 'pathological laboratory'. Although Dukes' professional qualifications in public health certainly marked him as much more open to social medicine, Lockhart Mummery was more interested in his statistical expertise. The possession of such expertise led him to entrust Dukes with the collection of the 'family histories' of patients suffering from 'polyposis intestini'. Importantly, Dukes realised that his expertise was, in fact, limited, and that knowledge of how to collect and analyse what were now understood as 'pedigrees', as opposed to 'family histories', rested with the Eugenics Society. Dukes eventually joined the Society, just as members of Medical Research Council's Committee on Human Genetics were seeking to shift the attention of the Eugenics Society to politically unproblematic clinical pathologies, hoping thereby to win greater acceptance of genetics within the medical profession. Dukes could help to advance this effort. His importance to geneticists was particularly evident in 1948, when he was asked to chair an international symposium jointly organised by the British Empire Cancer Campaign and the Genetical Society on 'the genetics of cancer'. Dukes was then the perfect candidate for the appointment as the director of the Research Department.

Dukes, thanks to his newly acquired position and the important place it occupied in the institutional organisation of St Mark's Hospital, could then begin to demand that his senior colleagues should collect and hand over to him blood samples from those patients who were afflicted by polyposis coli. These senior colleagues' proprietary relationship with their patients continued to create problems, but Dukes succeeded none the less in forwarding increasing numbers of blood samples to the Galton Laboratory, the leading British centre for research into the inheritance of human pathologies. Here, Lionel Penrose, one of the original members of the Medical Research

Council's Committee on Human Genetics, was busy breaking such research free from its association with eugenics by detaching interest in the demography of apparently heritable human diseases from questions of social policy. Dukes also sought to expand the 'pedigrees' collected in St Mark's Hospital by publicly calling on medical professionals around the world to inform him of any cases of polyposis coli they might encounter, to ascertain whether they were in fact cases of 'familial polyposis coli'. In other words, the investigation of this condition was no longer the pet project of a surgeon at St Mark's Hospital, intent on improving the efficacy of his surgical interventions, but was being relocated into the heart of the new and international field of 'human genetics'.

This relocation was accompanied by a much greater openness toward prevention, a central tenet of social medicine. Thus, in 1951, the first of an increasing number of more affluent patients to come within Dukes' purview expressed some concern about the reproductive implications of their condition, asking Dukes whether they should be sterilised to avoid bringing into the world similarly affected children. Dukes recommended that they should instead visit a 'good family planning clinic'.[12] Others, such Tom Rowntree, a young surgical registrar at St Mark's Hospital, were not at all averse to thinking about the more drastic actions sometimes envisioned by these patients, and publicly endorsed by Haldane. Arguably, however, this no longer constituted eugenic advice because it was not predicated on the social, if not racial, biases that once characterised such advice. Yet, the language Dukes used to describe how to collect family histories suggests otherwise. He wrote that,

> In each family one individual is selected who is called the collaborator. This person is chosen with care. . . . The essential quality of the collaborator must be that he or she is 'tribal' in outlook, is the sort of person who knows nephews and nieces or aunts and uncles. Having

chosen the collaborator, I make note each year in my diary of his or her birthday and write annually so that the birthday letter arrives on the right day. Before writing the letter I consult the family chart, making note of the members about whom information is most needed. Then after expressing the usual birthday greetings I inquire after little Alice or Sister Susie or Uncle Tom or whomever it may be, enclosing, of course, a stamped addressed envelope for reply.[13]

The connotations that the word 'collaborator' may have had in post-war Britain, especially when the patients often worried about 'informing' on their relatives, certainly were unfortunate. More importantly, however, the general tenor of these instructions was inflected by a 'anthropological' outlook, if not by an overtly racialised understanding of the lives of residents of the London East End. Such inflection was still, if not even more, evident in 1977, when Richard Bussey, Dukes' long-serving assistant, wrote to a colleague that,

We have sent our beaters out after some polyposis children who have not been seen for a while or not at all. One of these patients has apparently been caught in your net.[14]

From a practical perspective, social distinctions, grounded in the racial differentiations that once shaped the language of eugenics, including British eugenics, continued to inflect the language of human genetics.

Strikingly, the families on which the construction and extension of the Dukes' pedigrees depended often resisted Dukes' entreaties for information, and the correspondence in the files of Polyposis Registry now provided much more insight into the unwillingness to co-operate. In 1949, for example, Dukes' journey to visit one patient's relative, a 'labourer' in Blackburn, were repeatedly frustrated, largely because this relative protested that they were quite healthy and had no intention of submitting to a sigmoidal inspection.[15] Two years later, another patient was discouraged from collaborating because their sister, a nurse, warned them against becoming a 'guinia pig for surgions [sic]'.[16] More interestingly, these patients sometimes had very different ideas from Dukes and his colleagues about the reasons for their condition. Dukes' explanations that polyposis 'runs in the family' made no sense at all when 'Aunt Betty', who lived next door, died of cancer, even though her 'relatives' were now told that she was not 'really' part of the family. 'Black sheep' and illegitimate offspring, as well as different notions of kinship, often tripped Dukes and his colleagues, as they sought to link family histories and polyposis coli, and thus transform cancer into a genetically determined disease. Their difficulties sometimes led the patients to attribute the high incidence of cancer in their family to a shared history of bad diet, rather than to shared genes. Such refractory voices could not be ignored, without risking the loss of precious informers. Dukes and his colleagues then had to find a way of taking dietary factors into account and still sustain their preferred, genetic understanding. Thus, Hugh Lockhart Mummery, who had followed in his father's steps by also becoming a surgeon at St Mark's Hospital, suggested that Dukes should respond to a patient's dietary explanation by arguing that

> The dominant gene may be of low penetrance and that the hardships and intestinal upsets caused by his time as a prisoner of war in Japanese hands may have caused the appearance of overt disease.[17]

The tension between, on the one hand, Dukes' and his colleagues' commitment to a genetic explanation, and the patients' environmental account, on the other hand, was being resolved by mobilising the concept of 'genetic penetrance', an index of the extent to which a genetically determined condition is clinically manifest.

In sum, the genetic explanation of polyposis coli thrived under a National Health Service dedicated, at least in principle, to social medicine and the expansion of cognate bio-medical disciplines such as genetics. Its network expanded well beyond the walls of St Mark's Hospital. Following Michel Foucault, we might say that it became an integral part of a new and expansive 'disciplinary' apparatus. The National Health Service, however, was part of a reform of the British state that also encouraged once marginal people to oppose more brazenly the entreaties of their erstwhile social superiors. Gaining access to a specialist hospital such as St Mark's Hospital, for one, was no longer a business of appealing to the charitable instincts of the hospital's patrons, but a right. Patients expected to be treated on their own terms, not as 'guinia pig[s] [sic]', and sometimes they expressed their equality by articulating alternative explanations of disease, which had to be somehow confronted. Thus, as the genetic explanation of polyposis coli circulated more widely, it also became more open to destabilisation.

Enter the laboratory

One respondent to Cuthbert Dukes' call for records of patients afflicted by 'polyposis coli' was Arthur Veale, a clinician in the New Plymouth Hospital, in New Plymouth, New Zealand. As Dukes and Veale exchanged notes about the possible genealogical connections between some of the families in their respective registers, Veale became very interested in Lionel Penrose's work on the linkage between polyposis coli and genetically determined haematological markers. In 1960, he was appointed to a joint research post in St Mark's Hospital and the Galton Laboratory.

Veale, however, was not interested in the demographic questions explored by Penrose, but in the light that the genetic explanation of polyposis coli might shed on the process of carcinogenesis. Thus, in one of his first reports on the progress of his research in St Mark's Hospital and the Galton Laboratory he argued that,

> If the onset of malignancy at a particular site is determined by the completion of a 'partial' mutation, the existence of such a mutation could be proved by demonstrating that it was linked with some other genetically determined factor. This would contribute more to a theory of carcinogenesis than any number of associations or family studies.[18]

Once Veale was convinced that polyposis coli was an 'autosomal, dominant genetic disorder', the task was to understand why the 'FAP' mutation did not completely determine the onset of cancer. What this new task involved was statistically complex 'linkage analyses' to establish connections with other genetically determined loci, whose physiological and biochemical manifestations were better understood than was the case for polyposis coli. From now on, then, family records would no longer play a significant role in understanding the genetics of cancer. The British Empire Cancer Campaign, which had funded the collection of familial data in St Mark's Hospital for nearly forty years, ceased to do so. Richard Bussey was then appointed to transform the now financially independent Polyposis Registry into a reference collection for researchers well beyond the confines of St Mark's Hospital. It eventually became a reference collection for the World Health Organization.

It seemed that the future belonged to Veale and his intellectual successors, molecular geneticists. In the clinical ward, all that was needed to check for polyposis coli was a blood test for the increasingly refined biochemical signals derived from linkage analyses. John Northover, a surgeon at St Mark's Hospital, put the argument bluntly in 1984, when he introduced a new unit for the study of the molecular genetics of colorectal cancers. He stressed that,

> St. Mark's has played an important part in the evolution of the surgery of colorectal cancer, but surgery alone has reached its limits as a curative measure, and other methods of treatment must be

> explored. . . . New pathological techniques are being developed which reveal clinically important information on the biology of the disease, and these need to take their place in the assessment of patients at St Mark's.[19]

If the Polyposis Registry had any role in these novel developments, it was as a testing ground. In fact, the ready access it provided to a large population, whose genetic structure was understood much better than was the case with Georgiana Bonser's 'human subject[s]', meant that it was the ideal population for the clinical trial of the first chemical therapeutics that emerged from the research programme set in motion by Veale.

This downgrading of the Polyposis Registry from research tool to resource for therapeutic trials was, however, premature. The genealogical approach, seemingly exorcised by Veale's deft removal of the FAP mutation into the laboratory, continued to haunt the latter. The intensification of research into the molecular basis of cancer, which Veale effectively pioneered, was based on the assumption that familial polyposis coli was a homogeneous, genetically determined condition. Yet, the very effort to explain the incomplete 'penetrance' of the FAP mutation, by progressively excluding problematic cases, led to the proliferation of related, but none the less distinct, forms of polypal infestations of the colorectal tract. The uncertain genetic status of these excluded forms, however, was problematic for the grander significance of familial polyposis coli as a 'model' for the genetic determination of cancer more generally. Veale, for example, worried that these disruptive, anomalous cases were due simply to the notoriously incomplete information about the families listed in the Polyposis Registry. Such problems then called for a return to the Polyposis Registry. Strikingly, the difficulties confronting Veale bear a very close resemblance to those confronting the oncologists discussed by Ilana Löwy and Jean-Paul Gaudillière in their essay 'Disciplining cancer'. Such difficulties continue to this day, as 'modifier genes' have to be discovered to sustain the genetic determination of the newly re-named clinical condition 'adenomatous polyposis coli'. The important point is, however, that the genealogical constitution of the gene was not easily overcome, and the Polyposis Registry continued to be a crucially important, if now removed and invisible resource, which, coincidentally, began to expand more rapidly than ever before.

With the contemporary increasing democratisation of British society, the once less than aristocratic became important political actors. One aspect of this newly acquired importance was the growing popular interest in 'family trees', which, like social and local history, became an occasion for the celebration of heritage among people who once did not have a history. The Polyposis Registry contributed to the development of this 'history from below' by becoming a centre for exchange of family histories between patients and medical professionals, and, through them, between patients across the world. One particular patient discovered to their great excitement previously unknown relatives who had emigrated to New Zealand. In thus willingly participating in the construction of the family records, for reasons quite different from those of the medical professionals in the Polyposis Registry, the patients learned the new notions of kinship articulated by these researchers, and quite a bit about the genetic determination of disease as well. They began to speak with some confidence about 'genes', although Kay Neale, the Registrar of the Polyposis Registry, noted that they often got their 'mendelian ratios mixed up'.[20]

We need to be careful, however, about assuming that these patients had finally been incorporated into an emerging discourse of 'bio-power'. On the one hand, the increasingly more accurate identification of patients who were affected by polypal infestations of the colorectal tract, but not in the form of 'familial adenomatous polyposis', meant that the Polyposis Registry could ignore them.

They simply became as anonymous as any other patients in the clinical wards of St Mark's Hospital. One such patient complained about their relegation by reporting how they were told that

> The special polyposis clinic is a research clinic and you are not a suitable case for this! . . . What surprises me is that in earlier years – patients with polyposis, were always told that if there was any worries etc. don't hesitate to get in touch and the staff would help etc. It seems times have changed.[21]

In other words, such patients were no longer political subjects, intimately involved in the evolution of the genetic explanation of familial adenomatous polyposis, but were becoming instead the passive objects of disciplinary knowledge evoked by theories of professional, if not social, control. On the other hand, one might note how this patient's anger stemmed from their exclusion from the medical world, which could then be understood as indicating how this patient's identity had become tied to the medical domain, without any professional interference. As such, this patient's protestations would support Michel Foucault's arguments about the constitution of the subject of 'bio-power'. The meaning of any statement is ambiguous, to say the least.

The ambiguities of a family history

In 1992, prospective parents who were, or had at one time been, listed in the Polyposis Registry received a letter to inform them that,

> There have been some exciting new advances in our understanding of the genetic basis of polyposis, and this can provide us with new methods of . . . testing an unborn baby.[22]

This letter came from a new 'genetic counselling' clinic, whose establishment coincided with the growing financial difficulties confronting St Mark's Hospital in the wake of a renewed national effort to reorganise the provision of medical care.

The advent of the molecular markers presaged by Arthur Veale's investigations and eventually recommended by John Northover promised more a definite identification of those members of an affected family who did not carry the gene than could possibly be afforded by statistical calculus of mendelian genetics. For the increasingly prominent 'managers' at St Mark's Hospital, these molecular markers promised relief from the need to call in members of these families for periodic examinations, and thus very important financial savings. In other words, the new managerial rationality and genetics were close bed-fellows. It is perhaps not surprising then that the ghost of eugenics should have haunted Shirley Hodgson, the clinical geneticist heading the genetic counselling clinic. Like many of her colleagues, she worried that the counsel they offered might be construed as normative, and thus open them to accusations of renewing eugenics. She and others involved in this new approach to the management of polyposis then drew much comfort from the support for screening among their patients. As Northover pointed out, these patients were,

> Far more in favour of early diagnosis and application of linkage data to family affairs than their medical attendants might have thought.[23]

Not everyone, however, has been able to disengage the test and the ethical, if not political questions it raises as easily as Hodgson or Northover. Some patients have assumed that a positive diagnosis entails necessarily the termination of the developing foetus. One prospective parent, for example, declined to be tested because they 'did not think that a termination of pregnancy would be justified'.[24] Religious beliefs may have underlain this response, but it may have also been due to reading their family history differently from medical professionals. Rather than focusing, as the latter were wont to do, on the fate of any prospective children, many parents may have begun to focus instead on how they came to find themselves consulting these professionals. They may have begun to realise from their family histories, which they themselves helped to construct, but read into the past rather than the future, that, even if affected, their children could eventually be operated on and live a relatively normal life, just like their father or mother. For them, testing may have then been a matter of preparing themselves to live with an invalid child. As Hodgson reported after meeting a prospective parent, some of these patients were not worried by any 'psychological burden . . . of having an affected child'.[25] In other words, the genetic information the patients are increasingly receiving from the genetic counselling clinic is being translated into a personally meaningful datum to manage the risks and inevitable complications of everyday life, and sometimes, but only sometimes, do they opt for abortion. Of course, as Sir Walter Bodmer has noted, much to the alarm of the journalist who interviewed him, these choices come at some cost to the increasingly insolvent National Health Service. Thus, the British government is currently seeking to encourage the British public to opt for private insurance policies, so as to relieve the increasing financial pressures on the National Health Service. In the meantime, insurance companies are seeking the government's permission to use genetic tests for familial adenomatous polyposis, among seven key heritable diseases, to cost these insurance policies. It remains to be seen how the consequently increased cost of having a child who will most probably have to undergo surgical treatment to remove the pre-cancerous polyps that will infest his or her colorectal tract during teenage years will affect parents' decisions. They might want to think twice about the 'seed'.

In the meantime, one might begin to say that, when Theresa Marteau argued that the patients she studied, including members of the families in the Polyposis Registry, misunderstood the nature of their condition, she herself misunderstood the wide gulf between her and them. Puzzled by their responses to the offer of genetic screening, she suggested that

> Even though they have been attending clinics and know quite a bit about [the inherited nature of polyposis], they still conceptualise it as being a multi-factorial condition . . . it is not that they are ignorant, it's just that people have a sense that a gene may be necessary but not sufficient; there are environmental triggers. Scientifically, I don't think it's a bad way of thinking about it.[26]

Her response, however generous toward patients' understanding of genetics, rested on the sociobiological assumptions attendant on the contemporary expansion of genetics outside the laboratory. Marteau began with the gene as the foundational unit of analysis and admitted that environmental factors might sometimes mitigate its effects, though not in the case of familial adenomatous polyposis. For parents, however, belonging to a family carrying a 'problem' gene may only be one among many factors that will shape their children's life. It may be no more significant for the making of a good life than either these parents' religious values or the variety of services that might attenuate the suffering of these children. These services include the same surgical interventions in later life that put the parents in the position to weigh the relative merits of these different factors. From this perspective, familial adenomatous polyposis is indeed a 'multi-factorial condition', though not in the disciplinary, biological sense Marteau intended, which posited the environment as a disturbing rather than constitutive factor. Thus, even among those who share a common vocabulary, that of genetics, the appropriate way of reading a 'family history' is far from fixed, and thus a 'gene' is not always a 'gene'. Communities of knowledge, or at least linguistic communities, are not necessarily communities of practice, thus opening room for the production of more knowledge and new practices. As William Butler Yeats once put it, 'things fall apart; the centre cannot hold, mere anarchy is loosed upon the world'. [27] However, I would remove the adjective 'mere' and add that 'anarchy' is the norm, which then raises questions about the narrative of this chapter, if not *Plants, Patients and the Historian* as a whole.

Dealing with the incommensurable

The development of reproductive technologies such as those described in the report in *The Sunday Times*, with which this chapter started, is often viewed as a step toward a fundamental reconfiguration of human life. Some fear this reconfiguration as a return to eugenics, this time, conducive to logic of advanced consumer capitalism rather than the logic of the corporatist state. By focusing on the history of the genetic explanation of cancer on which the pre-implantation embryo screening that motivated the report in *The Sunday Times* rested, and reading against the grain the recording practices that produced this genetic explanation, I have argued instead that the situation is far more complex.

Percy Lockhart Mummery, the surgeon at St Mark's Hospital with whom I started, does not appear in any of the official histories of 'adenomatous polyposis coli'. These open instead with the work of Cuthbert Dukes, the first director of the Research Department at St Mark's Hospital. Lockhart Mummery's name does, however, appear in many of the files stored in the Polyposis Registry, a relatively invisible department within St Mark's Hospital that has proved critically important to the development of the pre-implantation embryo screening at the centre of the report in *The Sunday Times*. The recording practices that lie at the heart of the Registry's work set what Michel Foucault would have called 'the conditions of possibility' for the evolution of a eugenic understanding of familial adenomatous polyposis. Initially, the primary purpose of these practices was to reveal the interior of the clinical body, so as to improve the effectiveness of Lockhart Mummery's surgical interventions. What was required for the actualisation of the possibility of a eugenic understanding of familial adenomatous polyposis was the establishment of the National Health Service, which shifted the balance of power between the disciplinary practices of 'clinical' and 'social' medicine. Significantly, however, patients have been intimately involved with these developments because the extension of the Registry's records required their acquiescence to sharing information about their otherwise healthy families. Such acquiescence could never be taken for granted, especially in the wake of the transformation of British politics that led to the establishment of the National Health Service. These patients sometimes simply refused to collaborate in the genealogical investigations advanced by the personnel in the Polyposis Registry because they viewed medical professionals' enquiries about their relatives' health as unjustifiably intrusive. Occasionally, these patients' opposition has also taken the form of an explanation of their condition that is diametrically opposed to that proposed by the medical professionals. Most recently, however, these patients have argued instead that the genetic explanation of familial adenomatous polyposis only provides a partial answer to their specific situations. The meaning of a 'family history' is not univocal. Today it is instead weighed against personal considerations and the recollection of alternatives to the termination of a pregnancy, such as the surgery in later life that put the patients in the position to weigh the options. Maintaining the alignment of medical professionals' and patients' understandings of their condition has then been a business of constantly reformulating the genetic explanation, from the articulation of the concepts of 'genetic penetrance' and 'linkage' to a more sociobiological approach, in which the boundaries between genetics and sociology have become blurred. This conclusion should perhaps have been far from surprising since the science of genetics has always rested on the blurring of boundaries between the natural and the social, because genes are about 'networks of kinship'. I want to draw, however, some further, perhaps more debatable, conclusions. Knowledge never is commensurate with practice.

Contingency is one of the features of the narrative I have constructed around the report in *The Sunday Times*. The emergence of medical professionals' current understanding of familial adenomatous polyposis was the far from inevitable. The reason is that medical professionals do not encounter patients in the species existence imagined by modern medical discourse, but as individuated beings, 'Mr X or Mrs Y or Ms Z', located in particular places with particular memories. By their inevitable multiplicity and heterogeneity, these responses can easily destabilise medical professionals' claims to knowledge, to a knowledge that transcends the particularities of 'Mr X or Mrs Y or Ms Z'. This, in turn, can evoke, and has evoked, a reformulation of such knowledge, which entails the incorporation of new considerations to enrol the dissonant voices of 'Mr X or Mrs Y or Ms Z'. Admittedly, the disruptive acts of remembrance that impel this dynamic are mediated, if not constituted, by the very genealogical records that were central to the constitution of familial adenomatous polyposis. Yet, this attention to practices and contingency does not explain why some patients read their pedigrees backward, into the past, rather than forward, into the future. Similarly, for all the contingencies that have interrupted the development of the professional discourse of genetics, the interruption has never halted it, perhaps because the eugenic dream never disappeared. Sometimes it was a haunting presence, from which Shirley Hodgson and John Northover sought to distance themselves. Sometimes, however, this presence became more concrete. When Sir Walter Bodmer hinted at the need for social responsibility, he echoed the same notion of responsibility that lay at the heart of the injunction 'only healthy seed must be sown'. In other words, there has always been an interplay between the empirical world, to which the words 'practice' and 'contingent' speak, and something excessive, something that is not of the 'here and now'.

This brings me back to Foucault. While trying to clarify what he meant by 'discourse', Foucault characterised it as a 'space of dispersion', the historically contingent spatial distribution of knowledge and practices that establishes a particular subjectivity.[28] In his earliest works, such as *Madness and Civilization*, the relationship between knowledge and practice was a tense one. They were incommensurable, but related entities. The consequent tension opened the possibility for a causal account of historical transformation. This tension, however, also opened room for powerful philosophical critiques. The most incisive of these was perhaps Jacques Derrida's questioning, in 'Cogito and the history of madness', of Foucault's ability to stand outside history and thus provide a historical account of the transformations with which he was concerned. This relied on an implicit, and to Derrida, paradoxical, appeal to the very kind of historical meta-narratives Foucault sought to reject. The critique certainly drove Foucault to establish an increasingly tighter relationship between knowledge and practice, most notably in *Discipline and Punish*. He produced thereby a monolithic discourse, in which no practice or knowledge was outside a functionally closed discursive complex. It was then reasonable just to erase the differences between knowledge and practice, and to argue, as Foucault himself did in the first volume of *History of Sexuality*, that change testified to a semiotic space simultaneously unifying and disintegrating. However, being dissatisfied with the endless arguments on language, knowledge and power, the anthropologist Paul Rabinow takes up the frustrations to which Foucault eventually admitted, in the face of the 'relentless theorization of writing' forced upon him by Derrida.[29] He proposes to hypostasise practice, rather than language, as the locus of perpetual integration and disintegration.

In his essay on 'artificiality and enlightenment', Rabinow returns to the 'anthropological triad – life, labor and language', to argue that the contemporary transformation of human existence, life, which is being advanced by genetic engineering, is not predicated on transformation in the domain of language, but in the domain of labour. In this domain, an increasingly unstable knowledge is an effect rather than a cause of the ever more rapid movements and reconfigurations across the boundaries between the worlds of political argument, on the one hand, and scientific and technological innovation, on the other hand. The challenge of the day is no longer to answer questions about what is, nor is it to answer questions about what should be, but to act. The point is advanced further in *French DNA*, a genealogy of recent debates over the ownership of the information encoded in French patients' genome. Here, Rabinow argues that contemporary bio-medical researchers should not be viewed as 'knowledgeable', but as 'experimenters' in a world where the once secure distinctions between the natural and the social are no longer tenable. This fundamental change may explain why knowledge no longer appears to enjoy its earlier authority. As Jonathan Freedland, a columnist for the *Guardian* has put it, we are today waving 'goodbye to the oracle', to the expert who once claimed to know what the future holds in store for us. More importantly, as Madeleine Bunting, another columnist for the *Guardian*, has noted, as a consequence of the growing scepticism about knowledge and the momentous transformation we expect from the decoding of the human genome, we are left with 'no moral compass' against which to set our bearings.[30] In the context of such uncertainty, Rabinow advocates a new ethic of being to replace the 'hermeneutics of suspicion' that informs questions such as those concerning a possible return to eugenics or the commodification of life itself. Despite the disagreements with Bruno Latour, which Rabinow articulates in his essay 'Epochs, presents, events', this alternative comes very close to being a normative rendition of Latour's 'actor-network' theory. It calls for a 'will to experiment' that is much more open to the future and the heterogeneous than the insouciant 'will to knowledge', which greatly worried Foucault, in the first volume of *The History of Sexuality*.

I am not sure what to make of either Rabinow's or Foucault's latter-day reformulations of the relationship between knowledge and practice, both of which certainly evoke change, but no temporality. Since nothing is ever stable, in any more than the most anodyne manner, anything historically significant literally vanishes into meaninglessness. Ironically, even the concept of a past age when knowledge was hegemonic vanishes. There can be no event to divide the past from the future. Foucault himself was tacitly unwilling to accept such evacuation of 'discourse', which eventually resulted in his conceptually problematic, but none the less powerful appeal to 'bodies and pleasures' as the veritable grounds of politics, if not history.[31] As Judith Butler has noted perceptively, in her essay on 'Revisiting bodies and pleasures', this allowed Foucault to think about the future of humanity. I then want to resist the closure around either knowledge or practice. I want to preserve what Jacques Rancière calls 'incommensurability', an irreducible, structuring irruption that initiates, and indeed is the very precondition of political engagement and historical transformation.

Reacting to Foucault's ideas on the relationship between power and politics, Rancière turns to classical constitutions and analyses of politics. He argues that politics exists only when discourse confronts the violence in which it must necessarily be grounded, but which also is, by definition, always external to its domain of considerations. In other words, politics is only possible where there is an *aporia*, an impossibility of calculation. Similarly, the evolution of 'polyposis intestini' into the 'FAP mutation of the APC locus on chromosome 5q21' was built upon a founding violence. This was the bestowal of families from the London East End with membership of a new community, the universal community of the genetically endowed, although the difference that gave them a voice in this community was one to be erased. That is, the discourse of genetics endowed, and endows, the families in the Polyposis Registry with nothing but a negative quality: their 'problem' genes should be eliminated, if not re-engineered into 'unproblematic' ones. Like the *demos* of the classical *polis*, the families in the Polyposis Registry can then only bring 'contention' into this community. All they enjoy is the power to say 'no'. The situation is undoubtedly productive. 'Contention' has powered the transformation of 'polyposis intestini' into the FAP mutation of the APC locus on chromosome 5q21'. This transformation certainly is a cause for celebration among the contemporary *aristoï*, the virtuous of the classical *polis*, who seek to bring about the well-ordered, or, as Bodmer puts it in more resolutely modern terms, 'efficient' community. Yet, this transformation also threatens the dissolution of politics. The families in the Polyposis Registry, like the Association Française Contre les Myopathies considered by Rabinow, in *French DNA*, are beginning to constitute themselves as an 'interest' group, around their knowledge of a shared 'genetic' identity.[32] Something very precious could be lost, and not just the functionally productive agonism that results from incommensurability. Tom Shakespeare would appear to focus on exactly this point when he concludes his analysis of the threats presented by the contemporary medical applications of genetics by turning to Derrida and writing:

> Earlier versions of eugenics did not have a disability movement to deal with: it seems to me, despite the difficulties, we could adopt a Derridean slogan: Vive La Difference [sic]![33]

As Shakespeare notes, we need those endowed with 'problem' genes. They act as a reminder of the contingency and violence of any discursive formation. They act as a reminder that productivity rests on violence. Yet, as Shakespeare fails to note, as he elides the difference between 'difference' and 'differánce', differentiation is also about a world to come, a world in which difference will be no longer. The presence of those endowed with 'problem' genes then sustains a dialogue more significant than an engagement unable to effect anything but the further extension of discourse. It returns politics to its proper place because it speaks to a knowledge that the world might be otherwise than it is, that it might be redeemed. As Bernard Barataud, the charismatic leader of the Association Française Contre les Myopathies who plays a central role in the narrative of *French DNA*, once put it, 'without knowledge there is no hope'.[34]

At the time when I finished writing all this, I decided that I should leave it to others to establish whether the decision of the families in the Polyposis Registry to constitute themselves as a group marked the realisation of Rabinow's discourse of 'biosociality', or Foucault's discourse of 'bio-power'. This was work for sociologists and anthropologists. As a historian, it seemed to me that the more important problem was how to produce a historical narrative that avoided the denial of incommensurability to which Rancière spoke. The kind of emancipating history advocated by Mary Fissell, a history that aimed to return 'ownership of their bodies' to 'ordinary people', dangerously oversimplified hegemony. The danger was that, in seeking the realisation of this emancipating history in an archive that had somehow escaped the reach of modern medicine, the well-meaning, commonsensical 'just' social historian risked becoming a better inquisitor than the famous 'grand inquisitor' of Fyodor Dostoyevsky's *Brothers Karamazov*. He or she would bring 'ordinary' people's silence into account, an account that was not of their time or place. Yet, my own approach, which sought to avoid the search for the archive that had escaped the reach of modern medicine, was no less violent. No amount of reading the recording practices of modern medicine against the grain could undo the founding violence that brought patients at St Mark's Hospital, as well as their relatives, within the purview of the discourse of the 'gene'. For all my good intentions, these patients and their relatives remained, to use Luigi Pirandello's tantalising words in 'Six characters in search of an author', 'puppets instead of men'.[35] It then seemed to that a historiographical politics that was true to the incommensurable, one that was true to those emancipating aspirations of history that lay at the heart of Walter Benjamin's 'Theses on the philosophy of history', had to start somewhere else other than the archive. Returning to Lockhart Mummery and Lord Moynihan's shared ambivalence toward Sir Frederick Gowland Hopkins' vision of a truly 'scientific' medicine was perhaps an apt point from which to start again.

Notes

1 This chapter is based partly on Palladino, 'Speculations on cancer-free babies'; and 'Between knowledge and practice'.
2 Fissell, *Patients, Power and the Poor*, pp. 1 and 15.
3 Rogers, 'Doctors to create cancer-free babies', p. 24.
4 For an introduction to Martin Richards' work, see Marteau and Richards, *The Troubled Helix*.
5 Moynihan, 'The science of medicine', p. 779.
6 Lockhart Mummery, *The Origin of Cancer*, p. 135.
7 Lockhart Mummery, 'Prevention of cancer', p. 155.
8 Bodmer, as quoted in Meek, 'Gene test plea to cut cancer of bowel risk', p. 5.
9 Brasher, 'Clinical and social problems associated with familial intestinal polyposis', p. 789.
10 Foucault, *The Order of Things*, p. xxii.

11 Haldane, 'The prospects of eugenics', pp. 9–10.
12 Polyposis Registry: Family 22: C. E. Dukes to patient, 7 August 1951.
13 Dukes, 'Familial intestinal polyposis', pp. 2–3.
14 Polyposis Registry: Family 30: H. J. R. Bussey to M. Orr, 23 February 1977.
15 Polyposis Registry: Family 3: C. E. Dukes note, 29 December 1960.
16 Polyposis Registry: Family 22: Letter to C. E. Dukes, 5 August 1951.
17 Polyposis Registry: Family 44: H. E. Lockhart Mummery to C. E. Dukes, 5 May 1952.
18 Veale, 'Genetics, carcinogenesis, and family studies', p. 178.
19 Northover, 'Imperial Cancer Research Fund colorectal cancer unit', p. 53.
20 Kay Neale, personal communication, 27 June 1995.
21 Polyposis Registry: Family 53: Patient to S. Ritchie, 11 January 1984.
22 Polyposis Registry: Family 4: S. V. Hodgson to patients, 27 January 1992.
23 Northover, 'Imperial Cancer Research Fund colorectal cancer unit', p. 39.
24 Polyposis Registry: Family 4: S. V. Hodgson to J. Nicholls, 25 September 1992.
25 Polyposis Registry: Family 33: S. V. Hodgson, note, 22 November 1993.
26 Vines, 'Star of the big screen', p. 14.
27 Yeats, 'The second coming', p. 211.
28 Foucault, 'Politics and the study of discourse', p. 55.
29 Foucault, 'Truth and power', p. 127.
30 Bunting, 'Diving into the unknown', p. 17.
31 Foucault, *History of Sexuality*, p. 157.
32 Interview with members of family 14, 10 July 1997.
33 Shakespeare, 'Back to the future?', p. 34.
34 Barataud, as quoted in Rabinow, *French DNA*, p. 40.
35 Pirandello, 'Six characters in search of an author', p. 78.

6

Metaphor, desire and the historian

> The power of memory does not reside in its capacity to resurrect a situation or a feeling that actually existed, but is a constitutive act of the mind bound to its own present and oriented toward the future of its own elaboration.

Impure phenomena

According to Friedrich Nietzsche, classical scholars rightly understood historiography as a pedagogical exercise aiming to produce virtuous citizens.[1] If humans are as different from other objects of inquiry, as I suggested in the previous chapter, this seems to me to raise questions about how I, the historian, should behave toward that other human object of inquiry, the historical actor. There is an unsettling similarity between my historiographical practices thus far, and geneticists' own practices that fully justifies the use of the word 'object' rather than 'subject'. I have been deploying historical actors, who I have reconstructed by endless digging through archival repositories, to serve my own historiographical purposes, almost as if they had no agency in the making of historical recollection. Even when I sought to let once silenced actors 'speak', as I did in the last chapter, it was for the instrumental purpose of criticising other historians and their historiographical approaches. Like Sir Frederick Gowland Hopkins, if not like Georgiana Bonser, I approached these objects of inquiry, expecting them to answer 'yea' or 'nay' to questions that were not of their time or place. As we shall see, I had in fact completely ignored the full implications of Lord Moynihan's worries about the perversities of the human 'witness', which, as I mentioned in the last chapter, he summarised by arguing that

> The observer is . . . confronted not with phenomena as they really exist, but with impure phenomena in varying degree, adulterated by the mind or affected by the will or temperament of the patient.[2]

Humans are ambiguous and contradictory creatures. Their humanity resides in precisely such ambiguity and contradiction, which impels them continuously to ask questions, questions that are ultimately about what it means to be human. Significantly, Thomas Söderquist criticised the social constructionist approach to writing histories of science, technology and medicine by echoing Paul Rabinow in his pointing out that this approach is grounded in a 'hermeneutics of suspicion' that objectifies historical actors and surrenders any ambition to moral guidance. He wanted to replace this with a 'hermeneutics of edification', grounded in biographies that would focus on historical actors' existential struggles with ambiguity and contradiction. The criticism on which this proposal was based seemed to me important, especially as I increasingly engaged with Steven Woolgar's argument that social constructionism simply replaced the authority of nature evoked by natural scientists with the authority of society evoked by social scientists. Although Woolgar then sought an alternative science of science, technology and medicine, grounded in indifference to the traditional differentiation between the world of things and the world of people, I found his initial call for 'reflexivity' much more provocative. Once historiography, like science, is viewed as something more complex than an exercise in the objective representation of past events, I felt that it should begin to address more forthrightly the relationship between the observing historian and the observed historical subjects. It also seemed to me, however, that Thomas' 'hermeneutics of edification' was, in some ways, deeply problematic. As Thomas and I discussed his approach, I argued that, if it was to succeed, it had to address quite directly how Thomas himself was transformed by the engagement with his preferred historical subjects for, otherwise, these historical subjects would become the outward projections of himself. Returning to Luigi Pirandello's words in 'Six characters in search of an author', it seemed to me that, for all his good intentions, Thomas risked transforming his historical subjects into 'puppets instead of men'. It was at this point that Thomas directed me to Nietzsche's 'On the uses and disadvantages of history for life', and I then I fell under the spell of the following statement:

> I trust in youth that has led me aright when it now compels me to protest at the historical education of modern man, and when I demand that man should above all learn to live, and should employ history only in the service of the life he has learned to live.[3]

I began to realise then that what I wanted was to write a history of life, for life. I began to do so as I tried to come to terms with the increasingly unsettling figure of Percy Lockhart Mummery.

Remaking the world

As I suggested in the last two chapters, Percy Lockhart Mummery wanted to transform medicine into a science just as much as his sometime vitriolic critics Sir Walter Morley Fletcher and Sir Frederick Gowland Hopkins. In the last of these two chapters, I also hinted that his ambitions reached well beyond the confines of the clinic. The clinic was only a point of departure.

In *The Origin of Cancer*, Lockhart Mummery forewent in typically modern fashion any classical epigraphs, which David Cantor, in his essay on neo-hippocratism in inter-war Britain, has associated with conservative clinicians of the time. Lockhart Mummery stressed instead that he had arrived at his self-consciously 'modern' theory of carcinogenesis by deploying 'the method of solving problems dear to the authors of detective stories', and by then claiming that, thanks to this method, medicine was finally

> Getting down to the fundamental facts of life . . . necessary if we are able to satisfactorily control disease in human beings.[4]

Lockhart Mummery, furthermore, extended this modern outlook to the institutional organisation of medicine. He played a central role in establishing proctology as a new professional specialisation. While this played an important role in rescuing the declining reputation of St Mark's Hospital, by transforming it into a specialist hospital, the emphasis Lockhart Mummery placed on the treatment of colorectal cancers was particularly important. At the beginning of the twentieth century, cancer had acquired great public visibility, exciting charitable instincts, which Lockhart Mummery deftly exploited to establish the British Empire Cancer Campaign. With funding from the Campaign, he was then able to create the department of pathology so central to the development of his theory on the relationship between cancer and heredity. In 1937, he summarised the need for such institutional developments in an address to the American College of Surgeons, in which he argued that,

> Today the surgeon requires a regular team of assistants if he is to do first class work. Often elaborate tests have to be made requiring the assistance of skilled scientists, a physician's report on the functioning of the patient's various organs has to be obtained, biopsy specimens have to be reported upon by a skilled pathologist, and the surgeon has to rely upon a number of persons beside himself to

assist him in bringing the case to a successful conclusion; in other words, the best surgery now requires team work.[5]

The emphasis Lockhart Mummery thus placed on specialisation and team work had profound consequences for the patient. As I have already pointed out, the collection of the family histories on which he built his theory on the relationship between cancer and heredity relied on a network of consultants from across the nation, who had passed through St Mark's Hospital as registrars. There, they learned from Lockhart Mummery the new, collaborative mode of specialist practice, in which the patient became an exemplar of a shared population. At the same time, cancer became a disease of the social body, which entailed in turn a need for greater attention to the organisation and reproduction of contemporary society.

It is perhaps not surprising, then, that Lockhart Mummery, the surgeon who was interested in the hereditary basis of cancer, was an outspoken defender of Lord Dawson's controversial promotion of eugenics as a vehicle further to advance the latter's more famous plans for medical and social reform. Significantly, these plans would prove very influential in the evolution of the National Health Service. In 1935, Lockhart Mummery argued in the *British Medical Journal* that

> Human genetics must inevitably become the most important social and scientific problem of the next few decades, if the human race is to make any serious progress towards something better.[6]

Lockhart Mummery then described the path of such progress in *After Us, or the World as It Might Be.* In this collection of essays on the future of humanity, he argued that the most pressing problems of the day lay with the overcrowded city, and that they should be resolved by rebuilding the urban environment more rationally and by then directing its economy along corporatist lines. Ultimately, however, he believed that 'the causes of the troubles of the social life of the community are due chiefly to the faults of the members of the community itself, and not to institutions'.[7] These members were so faulty because,

> The improvements in medical knowledge of the last one hundred years has resulted in reducing the infant mortality from about 40 per cent of all births to about 6 per cent in England, and about the same in other civilised countries. The medical service of India has so reduced the mortality of plague, cholera, and dysentery, and other forms of disease, which used to decimate the natives, that the population of India has doubled within fifty years.
>
> Not only has medical science reduced the infant and adult mortality of almost all nations, but it has enabled a very large number of congenital abnormalities and mentally deficient persons to grow up and reproduce themselves, so that such abnormal persons have enormously increased in proportion to their normal brethren. This enormous increase which has taken place in the populations of civilised nations during the last fifty years is to no small extent due directly to improvements in medical efficiency.
>
> Even more serious is the fact that the poorer types of humanity, due to bad breeding, have been enabled to survive for the first time in history. As a result, they have seriously increased. Working with the very best intentions, the medical profession throughout the world is responsible for a most serious increase in the world population, which is causing serious embarrassment to all nations. It is responsible for the preservation, and consequent propagation, of a vast number of C3 citizens, who would never have survived early childhood had medical science remained as it was one hundred years ago and had not discovered how to preserve their lives.[8]

The residents of Lockhart Mummery's reconstructed cities would then be relieved from the 'troubles of social life' by allowing breeding only between men and women approximating to the ideal citizen. Any imperfect progeny would be cast unceremoniously over the walls of the new Sparta. He summarised these plans for the 'state control of parentage' by writing that,

> We are living in an age of great and very sudden changes, and human minds and institutions move very slowly in response to such changes. But one cannot doubt that, with the new knowledge that we are gaining of the ways in which heredity works and the importance of controlling breeding of our future citizens, a great measure of control by the State over the manner in which its future citizens are to be bred will come in the not distant future. That such interference with personal freedom of the citizen in his own most personal affairs will be bitterly resented at first is certain. A considerable measure of education will be necessary before the individual citizen will see the necessity and the wisdom of allowing the State to decide the kind of children that are to be bred. But State control over breeding of children will certainly come. It is indeed inevitable unless we are to allow the human race to return to barbarism and man revert to the status of a wild animal.[9]

Lockhart Mummery would appear to have been unsure about the power of education and the general will to self-improvement, however, for he also argued that the realisation of this dream would have to await

> Some kind of autocratic government which has the intelligence, the courage, and the power to enforce the necessary decrees for man's advantage.[10]

In sum, individual difference and autonomy were to be sacrificed by the transparent logic of genetics for the sake of a perfectly ordered and functioning social body. It is equally unsurprising that a reviewer for the *Times Literary Supplement* should then have accused Lockhart Mummery of wishing to remake the world into an imitation of 'Russia or Germany or Italy'.[11]

Punctum and aporia

As we have already seen, this casting of Percy Lockhart Mummery as an unrelenting moderniser has served me well in my engagement with the historiography of the conflict between medical scientists and clinicians. Clinicians' supposed conservatism, I was able to argue, is an artefact of historians' desire to produce a historicist explanation of conflicts such as that between Lord Moynihan and Sir Frederick Gowland Hopkins. As I sought out everything Lockhart Mummery ever published, however, I came across his collection of essays on natural history, *Nothing New Under the Sun*, and, at some point in my investigation that I can no longer recall I found myself particularly attracted by the frontispiece to this collection (see figure 3, p. 30).

While the aeroplane has held an important place in modern imagination, and as such fits with the portrayal of Lockhart Mummery as an unrelenting moderniser, the aeroplane in this frontispiece is a rather more complicated allegory. A one-eyed bird or fish, whose (mono)oculism recalls the opening scene of Ridley Scott's *Bladerunner* and perhaps symbolises the modern and destructive desire for omniscience, is superimposed on the aeroplane. This strange machine or organism seems, however, to have flown too close to the sun, and, like Icarus, is falling into the stormy sea below. The strangeness of this frontispiece is reinforced stylistically. An Arts and Crafts woodcut is used to illustrate that integration of the organic and the technological vehemently rejected by the Arts and Crafts movement. This frontispiece then began to evoke indistinct and disruptive thoughts about modernity and modernism.

I took it for granted that the defining characteristic of modernity was the institutional translation of Illuministic principles to reorganise all aspects of human experience and promote thereby a more humane way of life. Yet, the conjunction of modernity and violence during the twentieth century, amply confirmed by Lockhart Mummery's vision of the future of humanity, has cast serious doubts on the humanity of the Enlightenment. These doubts seemed to be reflected by the fall of the new Icarus. This frontispiece might then be associated with the modernist critique of the Enlightenment, but in a similarly contradictory manner modernism has sought to free the modern subject by celebrating its violence as the enactment of its freedom. I should then have simply refused both, and yet I could not deny that there was something sublime about the grand architectonic ambitions of the modernity and the aesthetic violence of modernism. So I found myself pondering the significance of this once invisible frontispiece, unsure about what exactly was the question or answer I was trying to formulate. It was the moment of *punctum* and *aporia*. This piercing frontispiece lay precariously at the intersection between the past and the future, and the only way I could move forward was to move backward and retrace my steps. I began to take stock of aspects of Lockhart Mummery that had previously gone unnoticed.

From transparency to opacity

There can be no more striking example of Percy Lockhart Mummery's greater complexity than I have been prepared to contemplate thus far than the same address to the American College of Surgeons in which he celebrated specialisation and team work. He also issued on this occasion an emphatic warning against 'becoming over specialised and too much like the machine minders of the mass production factory'.[12] Similar ambiguities are evident in the debates during the 1930s between Lockhart Mummery and William Gye. They contested incessantly the merits of their competing theories of the causes of cancer. They began by questioning the 'reality' of the equally invisible viruses and genes each proposed to account for the various characteristics of cancerous cells. They ended by disagreeing over the relative merits of Gye's 'experimental' method of inquiry and Lockhart Mummery's 'deductive' alternative, 'the method of solving problems dear to the authors of detective stories'.

These disagreements raised questions about the nature of scientific explanation that were so intractable that the editor of the *British Medical Journal* had to call a halt to the increasingly unproductive exchanges between Lockhart Mummery and Gye. Strikingly, these questions were already evident in Lockhart Mummery's introduction to *The Origin of Cancer*, when he tempered his profession of faith in the transparent logic of science by also writing that

> A theory may be described as a tentative framework or hypothesis to explain the known facts about a series of phenomena. The value of the theory is that it allows one to make deductions as to the probable sequence of causes and results and to arrange the ascertained facts in their proper order and importance . . . If there is any undoubted fact which cannot be fitted into the theory, then the theory must be discarded, or at least modified . . . Theories are always tentative, since new facts may be discovered which require very considerable modification of the original theory.[13]

Since Lockhart Mummery's belief in the fallibility and impermanence of human understanding went hand in hand with a deep conviction that the course of evolution was absolutely unpredictable, the notion that science could ever reflect an underlying reality was, to say the least, fanciful. Following Carlo Ginzburg, who argues that the appearance of the detective, specifically Arthur Conan Doyle's Sherlock Holmes, signals doubts about the transparence of modern reason, Lockhart Mummery's recommendation of the detective's methods against those of the experimentalist should then have signalled his greater affinity with modernism, than with modern optimism. In fact, the introduction to Lockhart Mummery's earliest, sustained discussion of cancer, in *Diseases of the Rectum and Colon*, suggests reservations about the impositions of contemporary society very close to those articulated by modernist critics of modernity. He inverted the usual relationship between civilisation and health, explaining all too resonantly that the growing incidence of cancer was due to modern, industrial methods of food manufacture. Specifically, he wrote that,

> In our present high state of civilisation . . . the animals which supply our meat are specially bred and cared for to render the meat free from gristle, and the vegetables are cultivated to contain but little cellulose, and are further prepared, especially in the case of bread, to reduce this ingredient to the very minimum. Under these conditions . . . the normal stimuli to peristalsis and digestion are to a large extent absent.[14]

Lockhart Mummery also suggested that the increased incidence of cancer was due to 'our disgusting habit of burying the dead', and concluded that this justified its replacement by cremation.[15] That is to say, the factory's chimney-stack, perhaps like that in Auschwitz, was the answer to the problems of modern civilisation.

In sum, the ambiguities, if not contradictions, of these statements reveal the traces of a dialogue between Lockhart Mummery, the figure I conjured in my imagination as a transparent symbol for the logic of science and the organisation of modern society, and something altogether different. This is what another historian might call, perhaps prematurely, the natural world, in the form of the cancerous cell, or the social world, in the form of modern society. More importantly, because Lockhart Mummery so resisted the logic which first impelled my initial historiographical mobilisation, I had to begin to contemplate that I too was confronting something similarly different. Perhaps, this is the voice of the historical subject to whom Thomas Söderquist had drawn my attention, but the question was then about what lessons could be taken from Lockhart Mummery' ambiguities and contradictions. This required that I weave some coherent narrative around these ambiguities and contradictions, so I returned to *The Origin of Cancer* to see how Lockhart Mummery managed his own encounter with alterity.

Order and autonomy

For Percy Lockhart Mummery, the fundamental task facing all scientific investigators interested in understanding the causes of cancer was to explain how balanced growth, the very antithesis of cancer, was established and maintained. He was wary of the hierarchical explanations proposed by the physiologists J. S. Haldane and L. J. Henderson, asserting that he found it simply impossible to 'imagine any central control mechanism which could bring about such a result'.[16] Instead he drew inspiration from the theory of the atom to write that

> The cell is an independent unit of life which is born, lives, and dies in its own place and time. Although dependent on other cells around for most of its activities and even for its continued existence, it has a separate individualistic life of its own.[17]

Sometimes Lockhart Mummery was even convinced that genes had to be viewed as the fundamental units of life, and that cancer was the result of a somatic mutation in individuals carrying unstable genes. He then explained historically why genes usually promoted orderly cellular proliferation and at other times determined that the cell should begin to proliferate uncontrollably. Specifically, he argued that the incidence of cancer was increasing among humans because evolution by natural selection was no longer operative, which then allowed populations exhibiting high rates of genetic mutation to survive. Among these surviving populations there were bound to be a greater number of individuals possessing genes for excessive cellular reproduction. Thus, thinking metaphorically, the 'dead' of *Diseases of the Rectum and Colon*, the progenitors of these unfortunate individuals, were indeed responsible for the increased incidence of cancer and cancer was quite literally the haunting disease of modern civilisation. It is then not surprising that Lockhart Mummery's contemporaries donated very generously to the British Empire Cancer Campaign. Yet, Lockhart Mummery's own use of metaphor in *The Origin of Cancer*, a scientific text in which metaphor should have no place, suggests that he was not completely satisfied by this historical explanation and exorcism. Combining the languages of nature and society, he wrote that

> The cancer cell may aptly be compared with the citizen of a community who having previously been a good citizen suddenly becomes a Communist and, believing in the destruction of all law and order, commences to live independently of his fellow citizens to his own advantage and to their detriment and destruction.[18]

Perhaps, then, the explanation of order did not lie in nature, in some intrinsic and essential nature of the cell, so let me turn to society.

The fundamental problem Lockhart Mummery sought to confront in *After Us* was to explain how the perfect society he imagined might emerge from his own, egoistically 'conservative' and sloppily 'sentimental' society.[19] Strangely, for someone who seemed to attach so much importance to individual autonomy and to some individuals' capacity to transcend their situation so as to be able to imagine a more orderly society, Lockhart Mummery was not prepared to appeal to genius. For all the authoritarian sympathies some attributed to him, he would not cast his lot with those notoriously brilliant minds who had radically transformed contemporary Italy, Germany, and Russia. Nor did he believe that eugenic planning would fulfil this need by providing more palatable figures, because mutation was such an unexplainable source of either destruction or improvement that planning the production of a new Charles Darwin was beyond all possibility. To realise his dream, Lockhart Mummery believed that a new religion, based in the history of human progress, was needed. This, however, was not the secular, evolutionary religion espoused by many of his contemporaries. Not only had this religion resulted in industrially manufactured foods and cancer, but it had also failed to provide a basis for commitment and right action. Humanity then needed a new godhead, and it would have to be 'Man'. Drawing on Winwood Reade's christological history of humanity, and mixing once again the languages of society and nature, Lockhart Mummery wrote that,

> Man's real God is not God in the attribute of man, but Man himself. Man's God should be the ideal man, not as he is now, but as he should be in the future. We live again in our descendants, whether they are our own creation, or other people. After all, what are individuals? We know that each of us is composed of a mass of millions of individual cells. Each of these cells is an individual living organism, which is part of a great community of cells, which together form a human being. In the same way millions of individual human beings go to make Man.[20]

Thus, for Lockhart Mummery, the future of humanity rested with freely thinking individuals, fully aware of humanity's history of trials and temptations, and of their duty to the continuation of this history. Sacrifice was crucially important. However, this new consciousness would begin to take shape only after a war more destructive than the First or the Second World War. Only after such a cathartic rite would

> Men and women become more worthy of their great place in nature than they are now. Their destiny has yet to be unfolded, but it will be toward the light – not towards darkness.[21]

These sentiments are much closer in spirit to the idealism that pervaded English philosophical circles during the last decades of the nineteenth century, and fostered liberal attempts to reconcile classic individualism and the emergent collectivist political discourse, than to the totalitarianism conjured by the reviewer for the *Times Literary Supplement.*

In sum, like all modernists, Lockhart Mummery was caught irreconcilably between his celebrations of modern notions of order and the uniqueness of the individual. Like the futurists, however, he sought to escape this dilemma by seeking refuge in the 'future', rather than in the 'past'. Thus, in *Nothing New under the Sun*, he wrote that,

> We now live in an age of bustle and hurry, of wars and rumours of wars, of insecurity of life and fortune, and it is perhaps natural that in our minds we should envy those ancestors of ours who lived in the time of Queen Elizabeth, of Charles the Second, or Queen Anne, when wars were only small affairs that did not personally concern the average citizen either directly or indirectly, and life went slowly and peacefully by. When there were no trains, motor cars, steamships or telephones; when bombs could not come down from the sky upon our streets and houses, and a war on the other side of the Channel was little more than a topic of conversation. I have often wondered how many of those who talk of the good old days have stopped to consider what the expression really means. Was the past better than the present as regards the things to which we attach the most importance? . . . No, the truth of it is that we have a lot to be thankful for and little to regret that we live in these times.[22]

More importantly, as I worked through these thoughts, I began to sense an unsettling parallel between Lockhart Mummery's struggles and my difficulties in making sense of who he was. We were both looking for coherence and transparence. Almost as if retracing the footsteps of Jules Michelet, who is today apparently enjoying a kind of revival, if we are to credit Raphael Samuel's last reflections on resurrection and contemporary culture, I then wanted to know how Lockhart Mummery lived outside the world of the word. Here, perhaps, I would find that for which I was searching. Enter Joseph Conrad's *Heart of Darkness.*

As I discovered in re-reading prefaces to Lockhart Mummery's writings, by digging in increasingly obscure and thinner archives, and interviewing Lockhart Mummery's surviving relatives, his favourite refuge from his professional responsibilities was the Savage

Club. The Club was famous for the iconoclasm of its members, drawn mostly from the world of literature, art, and drama. Landon Ronald, symphonic conductor, and composer of popular ballads, had invited Lockhart Mummery into the Club. Their friendship had been cemented by a shared passion for gambling, especially on Lockhart Mummery's prize-winning greyhounds. This lifestyle seems to have made Lockhart Mummery, the advocate of free love, immensely attractive to women, and his indiscretions resulted in a divorce loudly and damagingly publicised in the national press. As his daughter-in-law put it, it cost him the knighthood eventually earned by his son, Sir Hugh Lockhart Mummery. It is perhaps not surprising, then, that Matthew Stewart, Georgiana Bonser's mentor, described Lockhart Mummery as a raffish, bohemian man about town, with a penchant for the theatrical. Moreover, while Lockhart Mummery's watercolours did not look like anything that might be produced by a futurist, he was as enamoured as they were with the machine. It was from *Vie dell' Aria*, a stronghold of modernist aesthetics of the aeroplane, that he drew inspiration for the anti-aircraft projectile which he sought to patent at the beginning of the Second World War. Almost to close the links with futurism, I cannot resist adding that Lockhart Mummery viewed eating as a form of artistic performance remarkably similar to Tommaso Marinetti's dinners in the *Futurist Cookbook*.

All this so modern love for the novel and transgressive, however, did not stop Lockhart Mummery from acquiring an extensive farm, and enjoying showing it off to his bemused friends, all of which seems to suggest the need to pay attention to irony. In fact, Lockhart Mummery's close friend Lord Horder wrote in his forward for *After Us* that there were,

> Times when our minds . . . got some refreshment from a plunge into sheer irresponsibility. It was a form of mental gambling. But this writing of books is another matter, and . . . I was just a little shocked

> at the idea of your committing yourself to print. But you have safeguarded yourself by your frank avowal that you have been deliberately as violently controversial and provocative as possible, in order to try and make people think about the subjects with which you deal in a rational manner, unbiased by tradition and taboos. I don't suppose for a moment that this attitude will disarm the critics, but I know you well enough to believe that you will get quite as much fun out of their onslaught as you have got from writing the book.[23]

In the end, the playful, if not ironic, Lockhart Mummery fell prey to senile dementia. His life was eventually closed with one last act of irreverence, cremation.

Lockhart Mummery's life outside his writings on medicine, natural history, and the future of humanity, like his mutant and cancerous cell, subverted all accepted canons of proper conduct and defied all explanation. It was without rhyme or reason beyond itself. This then begs questions about the meaning of my effort to recollect and bring him back to life, about the meaning of a 'will to narrative'.

From writing about to writing with

According to historiographical conventions, this recollection of Percy Lockhart Mummery's ambivalence and contradictions is meaningful, that is, it contributes to the progress of historiography, only insofar as it relates to an established historiographical canon. This is what it means to be a historian, as opposed to what Friedrich Nietzsche called 'the antiquarian'.[24]

I could then have directed you, the readers of *Plants, Patients and the Historian*, to historians such as Chris Lawrence, who has argued that the institutional position of medical figures such as Lockhart Mummery or his more famous friend Lord Horder was threatened by the modernisation of medicine. Such modernisation called upon these figures to submerge their identity into that of teams of medical experts working on ever more anonymous patients. Not surprisingly, they then sought to resist such modernisation, even though, paradoxically, they did much to advance it. I could have then added that modernism, perhaps like Lockhart Mummery's iconoclasm, has been explained by Zygmunt Bauman, in *Modernity and Ambivalence*, as an aesthetic reaction to the turning of the modern gaze onto the inquiring subject, attempting to transform him or her into a further object of modern discourse. Returning to the frontispiece of *Nothing New Under the Sun*, the irony of Lockhart Mummery's situation may not have escaped him, even as he celebrated the accomplishments of science and modern civilisation. The new Icarus is crashing into the sea. Attending to such irony would have added to our understanding of modernity by giving us some insight into what Raymond Williams has called the 'subjective structures of feeling' about being a modern, alienated subject. In sum, I should have argued that historicising Lockhart Mummery brought some further order and transparence to events that have shaped profoundly who we are today. Irony is the modern condition.

Such historicism, however, would deny agency to all historical actors, as their voices are lost in the details of institutional organisation and politics. Only the historian is somehow allowed to rise above the deafening din. Some historians might then have preferred to this reproduction of the alienation of the modern subject that I engage instead with Lockhart Mummery as a subject, as such, or, as Thomas Söderquist has put it, with the 'subject as a human being'.[25] For the purpose of moral edification, I should then have concluded that Lockhart Mummery sought in his anarchic lifestyle an escape from an irresolvable tension. This was the tension between Lockhart Mummery's desire for order, implied by his participation in the objectivising discourse of modernity, and maintaining his autonomy within the world constituted by this very discourse. He dreamt of flying to the sun by being absolutely rational, to become his own God, only to fall back into the sea like Icarus and reap his just deserts: Lockhart Mummery lost his mind and was no more than the barbarians he feared. Lockhart Mummery's life could then stand as a parable about the good life of modernity. Oddly, the Raphael Samuel of 'People with stars in their eyes' would have enjoyed such a reading.

Like the modernists, however, Lockhart Mummery never spoke or wrote about himself with any reproach for his supposed ambiguities and contradictions. Ambiguity and contradiction is in the eyes of the beholder. The centring of the subject and moral reading proposed by Thomas would then be just as monological as any historicist explanation. Much more importantly, however, neither of the proposed readings provides any justification for their shared monological form. They both seem to be driven by an implicit desire to close the most vexing problem of modernity, the relationship between individual agency and social structure. This desire speaks incoherently of death. One denies agency altogether, by evacuating the subject, and the other seeks to re-establish agency through an identification with the subject that is in fact no less violent than the former. As Megan Boler has argued, ethical relationships, such as that sought by Thomas, cannot rest on 'empathy', as this can easily degenerate into 'substitution', but must instead rest on an awareness of unbridgeable difference, rather than a functional relationship between 'self' and 'other'.

I then wanted to focus on the strange power of Lockhart Mummery to create something new, or, if the term is still meaningful, Lockhart Mummery's power to make new subjects. Our practices of representation, however, fail me here, just as they failed Lockhart Mummery before me, so I will turn to Jacques Derrida's concept of 'writing under erasure' to clarify the thought on which I want to hold our attention.

As Derrida has argued in *Grammatology*, G. F. W. Hegel once noted that the 'preface', by setting out the question addressed by the text it prefaces, suggests that the question precedes the argument, when in fact it is a summary re-reading and re-presentation of the text. In this sense, the question always already has its answer, and what is philosophically important is instead the process of reaching for the answer, as it unfolds in the process of writing. Derrida extends the critique by adding that the preface and the text are related by repetition, but they are not the same. Although the 'preface' purports to reiterate the argument of the text, insofar as it offers a re-presentation, it is in fact not identical. At the very least it is a product of the process of reading and coming to see the text with a new and different understanding, which then means that recovering the original meaning of a text, the aim of every act of reading, is impossible. Derrida is not interested in such recovery, however, but in the productivity of the consequent play of identity and difference. The phrase 'writing under erasure' speaks to this same play of identity and difference, as it raises questions about, for example, the word 'I' that recurs throughout *Plants, Patients and the Historian.* Thus, if history has often been conceived as a teacher and that herein lies the political or moral function of historiography, I would claim instead that history is a mirror, and that even this analogy is unsatisfactory. I, the author of *Plants, Patients and the Historian*, and the mirror, the collection of fragments that goes by the name of Lockhart Mummery, did not pre-exist the moment of reflection. Lockhart Mummery and I have instead emerged together, in a mutually constitutive process.

As the death of ideology, history and society were first announced, a now distant historian literally constructed a 'puppet', Lockhart Mummery. Lockhart Mummery's biography, this historian argued, could pose difficult questions for the historicist explanation of conflicts between reforming medical scientists and conservative clinicians. It was a gambit to establish his own agency. In the course of this enterprise, that distant historian then began to worry about how historians piece together those archival fragments by which they define themselves professionally. Historians collect these fragments interminably to capture the reality of the past and its subjects, and yet they treat historical actors' accounts and their own explanations very differently. Historical actors' accounts are always dialogical, while the historian's are monological. The historian is the only subject. The struggle with this asymmetry, especially as the archival fragments this historian collected spoke uncannily about the same motivating problem, namely, establishing the place of the individual in the landscape of modernity, forced this distant historian to confront the destructive project of modern historiography. It reproduces alienation. Consuming, if not consumed by, the de-centred narratives of post-human and post-colonial fiction, from William Gibson's *Neuromancer* to Orhan Pamuk's *White Castle*, he began to realise that his own identity and that of Lockhart Mummery, this figment of the historiographical imagination, were inextricably tied. As this imagined figure resisted this historian's will to deny the fundamental, constituting irresolution of life, this figure became, as Luigi Pirandello put it, 'a living being with more genuine life than people have who breathe and wear clothes'.[26] At the same time, a new narrator began to emerge, one more willing to acknowledge the problematic nature of the self by translating these questions about modernity into new modes of historical narrative. Through an open dialogue with this figure, who is magically both son and father, I was coming to understand what it might mean to be concerned with being here and not nostalgic for lost worlds. As Paul De Man wrote some years ago,

> The power of memory does not reside in its capacity to resurrect a situation or a feeling that actually existed, but is a constitutive act of the mind bound to its own present and oriented toward the future of its own elaboration.[27]

Yet, this so transforming process originated with a question regarding the historian's proper posture toward the historical actor in the context of a critical analysis of debates between laboratory workers and clinicians over the modernisation of medicine. What was at stake in these debates was the medical investigator's proper posture toward the human subject, a debate that goes to the heart of both the subjective and sociological questions raised by the advent of 'designer babies' and the debates over the 'future of the human'. Lurking behind the analogy then drawn between the historian and the clinician was a prior assumption that the historian was also engaging with another human subject, when, in fact, all that the historian was engaging with were archival artefacts, and pieces of paper at that. This assumption and imaginative leap spoke of the same problematic humanism that impelled Thomas' search for a 'hermeneutics of edification'. As Mick Dillon has noted in his essay on 'another justice', which he suggested that I should read after he, Bruno Latour and I had been arguing over the merits of Latour's notion of a 'parliament of things', this grounding of ethical relationships entails a dangerously calculating rationality.

The 'parliament of things', especially as articulated in *Politiques de la Nature*, represents the extension of Latour's rejection of the categorical distinction between the social and the material worlds into the domain of political theory and ethics. Evoking environmentalists' search for a new relationship to the natural world, he couches this rejection in terms of an ethics of responsibility to the 'other'. Yet, his approach is grounded in Carl Schmitt's theory of political order. According to Schmitt, every action is the result of a calculating partition between 'friend' and 'foe' that is always grounded in some unacknowledged founding violence regarding what counts and what does not count. At one point or another, such calculation will therefore yield zero as its result, and then give way to the genocidal madness of the Holocaust. Drawing on, but also disagreeing with, the ideas of Emmanuel Levinas, Mick argues that an ethical relationship must then be predicated on an understanding of the 'other' that defies any kind of calculation whatsoever. The 'other' of the ethical encounter must become the 'Other', that which is so irreducibly different and unknowable that it constantly calls into question the referent of language, desire and the concern for justice that is central to the constitution of the self. In other words, ethics must always precede ontology. As Mick has also noted, however, if there is a radical, irreducible disjunction between the self and the Other, as Levinas claims, desire for the Other is then absolute, which raises questions about how to avoid self-annihilation. Mick's answer is to turn to the concept of 'hybridity', and argue that the Other is not external, but always already integral to the self.

It was at this point that I began to take rather more seriously the historian's engagement with the archive, understood simply as a repository of verbal, visual, auditory and tactile artefacts. I did so by re-considering what I had written about the history of agricultural genetics, and the relationship between agricultural geneticists and the plants they studied.

Notes

1 This chapter is based partly on Palladino, 'Icarus' flight'.
2 Moynihan, 'The science of medicine', p. 779.
3 Nietzsche, 'On the uses and disadvantages of history for life', p. 116.
4 Lockhart Mummery, *The Origin of Cancer*, p. 8.
5 Lockhart Mummery, 'The surgeon as a biologist', p. 259.

6 Lockhart Mummery, 'Medical science and social progress', p. 1022.
7 Lockhart Mummery, *After Us*, p. 222.
8 *Ibid.*, pp. 249–50.
9 *Ibid.*, p. 47.
10 *Ibid.*, p. 220.
11 Anonymous, 'A dream of AD 2456', p. 102.
12 Lockhart Mummery, 'The surgeon as a biologist', p. 258.
13 Lockhart Mummery, *The Origin of Cancer*, p. 7.
14 Lockhart Mummery, *Diseases of the Rectum and Colon and their Surgical Treatment*, pp. vii–viii.
15 *Ibid.*, p. 688.
16 Lockhart Mummery, *The Origin of Cancer*, p. 24.
17 Lockhart Mummery, *Nothing New under the Sun*, p. 129.
18 Lockhart Mummery, *The Origin of Cancer*, p. 1.
19 Lockhart Mummery, *After Us*, p. 274.
20 *Ibid.*, p. 148.
21 *Ibid.*, p. 281.
22 Lockhart Mummery, *Nothing New under the Sun*, pp. 103 and 110.
23 Lockhart Mummery, *After Us*, p. 15.
24 Nietzsche, 'On the uses and disadvantages of history for life', p. 67.
25 Söderquist, 'Existential projects and existential choice in science', p. 64.
26 Pirandello, 'Six characters in search of an author', p. 78.
27 De Man, *Blindness and Insight*, p. 92.

7

Writing and the experimental life

> The bourgeoisie cannot exist without constantly revolutionising the instruments of production, and thereby the relations of production, and with them the whole relations of society. Conservation of the old modes of production in unaltered form was, on the contrary, the first condition of existence for all earlier industrial classes. Constant revolutionising of production, uninterrupted disturbance of all social conditions, everlasting uncertainty and agitation distinguish the bourgeois epoch from all earlier ones. All fixed, fast-frozen relations, with their train of ancient prejudices and opinions, are swept away, all new-formed ones become antiquated before they can ossify. All that is solid melts into air, all that is holy is profaned, and man is at last compelled to face with sober senses, his real conditions of life, and his relations with his kind.

Being and the world of things

The process of historiographical production discussed in the last chapter began to raise questions about the relationship between the historian and archival fragments such as the frontispiece to *Nothing New Under the Sun*. These questions, however, lost their disruptive potency with respect to the traditional historiographical centring of the subject once the claim for the reciprocal constitution of the historian and the historical actor was revealed as grounded in prior knowledge that these fragments actually were traces left by Percy Lockhart Mummery. Lockhart Mummery, the forgotten historical actor, once was a human subject of flesh and blood, just like the historian who remembered his dismembered and scattered body out of a sense of charity, if not duty, toward a fellow human being. An ethics that is adequate to the Holocaust should, however, be grounded in the rejection of all such commensurability. The consequent task before me became increasingly evident as Tiago Moreira, who was working on material practices and the constitution of the bio-medical subject, introduced me to the poetry of Francis Ponge.[1] Significantly, Tiago's work, an extension of the Latourian rejection of any categorical distinction between the social and the natural worlds, signals a return to pre-Socratic philosophy, and specifically to Heraclitus. The aim is to escape the problems inaugurated by the classical distinction between the *bios* and *zoē*. Tiago's reference to Ponge's analysis of the relationship between being and the material world struck me particularly, as Ponge summarised this relationship by writing that

> La variété des choses est en réalité ce qui me construit. Voici ce que je veux dire: leur variété me construit, me permettrait d'exister dans le silence même. Comme le lieu autour duquel elles existent. Mais par rapport à l'une d'elles seulement, eu égard à chacune d'elles en particulier, si je n'en considère qu'un, je disparais: elle m'annihile. Et si elle n'est que mon prétexte, ma raison d'être, s'il faut donc que j'existe, à partir d'elle, ce ne sera, ce ne pourra être que par une certaine création de ma part à son propos. Quelle création? Le texte.
>
> (The variety of things is really that which constructs me. This is what I mean: their variety makes me, allows me to exist within silence itself. Like the place around which they exist. But in relationship to one thing alone, having regard to each of them in their particulars, if I consider all but one of them it annihilates me. And, if it is my pretext, my reason for being, if it is responsible for, and the root cause

of my existence, this is, this can only be thanks to a certain creation on my part regarding this thing. What creation? The text.)[2]

I began to wonder whether it was possible to have a subjective relationship with the world of things such as the verbal, visual or auditory materials, which the historian calls 'archival documents', in themselves and for themselves.

I had been fascinated by the difference between agricultural and medical researchers' relationship to their distinctive objects of inquiry. It was easy to imagine that the latter might be much more critical of the ambitions of modern science because their objects were human beings rather than silent plants or animals. In other words, I assumed that, unlike plants, patients were endowed with the capacity to resist those ambitions and thus elicit contradiction and ambivalence, the stuff of subjectivity. Now, however, I began to notice a common trope of the obituaries commemorating the agricultural researchers I had studied. These obituaries celebrated not just the researchers' contributions to the science of genetics and the agricultural economy, but also their aesthetic relationship to the world of plants. Sir Rowland Biffen's students and colleagues, for example, recalled how Biffen's 'heart was in his work, his garden, his art [and] the countryside', and how Biffen used to 'see a flower with an artist's eye'.[3] Perhaps these were not simply artistic liberties, disguising the agricultural geneticist's calculating rationality, but expressions of a subjective dimension of being an agricultural geneticist. As Evelyn Fox Keller has noted, in her acclaimed biography of Barbara McClintock, 'a feeling for the organism', specifically a feeling for the humble plant of maize next to which McClintock liked to be photographed, played a crucial role in McClintock's pioneering work on genetic regulation. Perhaps I had passed all too quickly over Redcliffe Salaman's *The History and Social Influence of the Potato*. Like the work of the first social historians, *The History and Social Influence of the Potato* speaks of love for the dispossessed, including the humble potato.

Inspired by Ponge's poetry, I then turned my attention to an old book sitting uncut and unread on my bookshelf, Sir George Stapledon's *The Way of the Land*, a collection of essays Stapledon had written and published over the twenty plus years during which he was the director of the Welsh Plant Breeding Station.

Refashioning the rural world

As is clear from *The Way of the Land*, its author, Sir George Stapledon, viewed plant breeding and the science of genetics as fundamentally important to the reorganisation of rural Britain.[4] British agriculture had been in such a depressed state since the repeal of the Corn Laws that nearly a third of the agricultural land across the country was under indifferent grass and scrub, either because it had never been actively cultivated or because it had been abandoned as unremunerative. For a nation that had faced food shortages during the First World War, and, in 1941, the year in which *The Way of the Land* was published, confronted both renewed food shortages and the prospect of their continuation long after the newest global conflict, this situation was simply unacceptable. Remedying the problem had to begin with the improvement of the land under grass by developing seed varieties that would turn this forgotten third of the land into something as productive as the arable lands in the Midlands and East Anglia. As Stapledon put it, 'seeds are the basis of farming'.[5] At the same time, however, he also felt that the geneticist's assumption that a plant's potential productivity was solely determined by its genetic characteristics was untenable. As he made it quite clear throughout *The Way of the Land*, Stapledon was in no way sceptical about the science of genetics, but he asked farmers and agricultural scientists,

> What is going to be the competitive interaction between species you sow? How is it all going to be affected by those indigenous species which have the knack of springing up naturally? And, furthermore, what are the antecedents of the seed of the desired species you sow? – where has the seed come from, and is it really going to succeed and fill up its allotted portion of ground? . . . Before we can employ mendelian methods or adopt the more correct and accurate plans of selection, we must know a great deal more about our indigenous plants . . . I think this assumption is not opposed to modern mendelian teaching, for the characteristics are in many cases 'strong potentialities' rather than absolute characters. [6]

Stapledon believed that a plant's successful establishment in a new environment, chosen either by the plant breeder or the farmer sowing its seed, was determined by the plant's place of origin and its history of adaptation to that original environment. Though this might seem to echo John Percival and Montagu Drummond's emphasis on taxonomic investigations, he was in fact sceptical of such investigations and much more interested in the 'ecological–genetical' approach pioneered by his 'friend' Göte Turesson.[7] That is to say, Stapledon was more interested in understanding the competitive dynamics between different species and how these could be modified by the introduction of new ones.

This said, Stapledon also pointed out that he differed with Turesson insofar as his goal was to improve the productivity of farmers' fields, and thus he was more interested in the competitive dynamics between different crop varieties rather than different botanical species. This arguably unorthodox distinction was based on the realisation that the dynamics of the former class were influenced quite profoundly by human interventions. Stapledon's lengthy, and often tedious, discussions of different types of soil and how they were affected by different fertilising regimes attest to the importance he attached to this human role.

It is in this conceptual context that Stapledon strongly recommended to his readers the science of 'agronomy'.[8] It was a way of thinking that, as he reported, was widely explored in the United States, and was immensely more valuable than the usual understanding of agricultural science as simply the practical application of conventional academic disciplines, which presumably was the norm in his own country. Moreover, while Stapledon never used the term 'inter-disciplinarity', it was exactly for this that he also called when he wrote that

> What is ideally wanted is to send a highly efficient team of chemists, biologists, geneticists, physicists, economists, and engineers, who know nothing about the canons of good husbandry, to a land that has never been cultivated, and to say to them: get busy, use your scientific principles, your scientific acumen, your common sense, and establish your canons of good husbandry, your rotations, your methods, strictly on the basis of exact experimentation.[9]

For Stapledon, 'free commerce in facts . . . [and] . . . ideas', blind to disciplinary 'nationalism', was the key to advancing a new vision of agricultural research.[10] Importantly, these interdisciplinary teams' healthy ignorance of the 'canons of good husbandry' also commanded ignorance of the canonical boundaries of a farmer's field. These teams were to adopt a regional, if not global perspective, both in terms of the problems they confronted and the knowledge on which they drew.

'Free commerce', more conventionally understood, was, however, the source of the problems confronting agriculture. If Britain was incapable of feeding itself and thus dangerously exposed in times of crisis, Stapledon argued, this was due to its having sacrificed agriculture for the sake of industrial growth and imperial security through 'free trade'. He claimed, furthermore, that the current belief that this state of affairs was somehow inevitable was grounded in ignorance, this time understood in all its negative connotations:

> We have more moral, political, and historical wisdom than we know how to reduce into practice – we have more scientific and economical knowledge than can be accommodated to the just distribution of the produce which it multiplies.[11]

After considering the environmental cost of overproduction in the countries from which Britain imported its food, to the cost of British farmers' economic well-being, he added still more bitingly that,

> The possibilities of generously feeding the whole of mankind . . . are adversely affected only by man's wilfully restricted capacity for breaking down the artificial barriers he himself set up by exaggerated deference to the free play of 'economic' laws. 'Laws' which are not Nature's laws and are therefore within man's own power to revise or repeal whenever he can summon enough courage or enough wisdom to revise or repeal them.[12]

For Stapledon, instilling such 'courage' and 'wisdom' to reshape the world, in a way that was more commensurate with the truths revealed by science, began with the refashioning the world's farmers.

In Wales, refashioning farmers meant convincing them to abandon Welsh culture. As Stapledon put it, 'although I am a naturalized Welshman of eighteen years' standing . . . I have a very grave suspicion that your language is not a good one in which to think accurately and scientifically'.[13] English culture, especially that aspect of English culture represented by Charles Darwin, was better suited to their needs. This extraordinary transformation could be advanced by quite ordinary means, such as involving the benighted Welsh farmers in farm trials of novel sowing patterns, even into the long abandoned highlands, all, of course, under the supervision of agricultural extension advisors. Importantly, Stapledon did not believe that Welsh farmers were exceptionally retrograde, and thus he hoped that the Ministry of Agriculture and Fisheries would eventually aid the extension of his programme beyond the confines of Wales. Every British farmer had to realise that

> The extent to which his Empire competitors have created their own markets by supplying what are tantamount to proprietary articles of reasonably good quality and always uniform standard. The one hope for the British farmer is to produce proprietary articles of standard quality higher than that of his Empire competitors . . . I would like to see the British farmer helped in every conceivable way; by research, by organization, by the setting up of central abattoirs and freezing works, by a more regularized system of regionalization in marketing, and by a regionalized system of commodity production to a fixed standard.[14]

It was not just farmers, however, who had to be educated into a more scientifically rational way of thinking. Referring to existing forms of standardised production, he wrote plaintively that,

> It is for the most part trade requirements – based on who knows what subtle psychological interactions – that have set the standards of quality. The millers prefer a certain type of wheat because they reckon the public like a certain type of bread. The vendors of fish and chips like a potato of certain shape and size because it chips well – and so it is on considerations such as these that the farmer is compelled to take his cue as to standards of quality. We, the plant breeders and the farmers behind us, are standing by, eager to be influenced by the researches of the human and animal nutritionists the world over and hoping to be told that a Cox's Pippin keeps the doctor away better than any other apple because of its tastiness, but willing to breed for any characters in our crops that are in the best interest of, and are demanded by, a super-enlightened people.[15]

In sum, Stapledon's vision of the future of food production and consumption was profoundly technocratic, as is amply confirmed by his statement that, 'the plant breeder confounds mankind, marching as he does ahead of the economist and the politician. He has already produced results to which the world at large has not yet contrived to adjust itself.'[16] The further, imperialist thrust of this vision was confirmed equally amply by Stapledon's recommendation that the measures envisioned by the Colonial Office for the transformation of agriculture in the African colonies and dominions should be adopted at home as well. These measures integrated every activity, from seeding to the marketing of the finished product, under the expert guidance of officers in charge of 'colonial development'.

As is again clear throughout *The Way of the Land*, the technocratic and imperialist vision it set out was not the stuff of utopian fantasy. Stapledon, the director of the Welsh Plant Breeding Station, and his allies in the Empire Marketing Board and the Imperial Bureau of Agriculture, had already begun to implement much of what Stapledon, the author of *The Way of the Land*, recommended to his readers. Significantly, after the Second World War, the Agricultural Research Council and the Ministry of Agriculture and Fisheries implemented this vision still further, though more with an eye to the corporate alliances that supported the Plant Breeding Institute, than to the National Farmers Union, to whom Stapledon often addressed himself. According to Norman Simmonds, one of the last directors of the Scottish Plant Breeding Station and, before that, a plant geneticist in the Imperial College of Tropical Agriculture, if Stapledon's vision was not implemented to the letter, it was because there was something far too 'mystical' about Stapledon.[17]

Modernist sensibilities

Sir George Stapledon's hero was the engineer, an engineer with the largest remit imaginable. As he put it in *The Way of the Land*,

> Where the land surface of this country is at stake we have an engineering problem of the first magnitude. There has been so much un-coordinated and unpremeditated action – both anathema to the engineer – that one fears that those of us who at once love the land

and are imbued with the spirit of the engineer are already fighting for a lost cause.[18]

Yet, Stapledon's understanding of how this engineer could resolve the problems of the agricultural economy suggests a more than simply technocratic and imperialist perspective. Besides hinting at the necessity for 'love' of the land, he also wrote that

> The engineer and the man of action are concerned always much more with fundamentals than with ultimates. Nothing is perhaps more surprising than the engineering feats that are possible and have been accomplished without any profound knowledge of the sciences, but always the fundamentals must have been very clearly understood. Any act of man conducted for the benefit and glorification of man can, however, only be fully successful if amongst the fundamentals which the engineer and the architect, equally with the social reformer, take into consideration are the attributes of man himself.[19]

Stapledon's celebration of the engineer who 'loved' the land and took into consideration the 'attributes of man' signalled his interest in William Blake's Romantic sensibilities and Henri Bergson's modernist vision of the world as being in a state of perpetual change. He appealed to Blake in his recommendation to prospective scientists and farmers that 'the man who never alters his opinion is like standing water and breeds reptiles of the mind'.[20] Following Bergson, he added that this healthier, fluid attitude of mind was necessary because scientists and farmers' object of study, nature, was itself a dynamic unit shaped by perpetual, nearly 'suicidal' strife.[21] This said, nature was only nearly suicidal because, *contra* Western metaphysics, life, rather than death, was the truth of all existence.

The 'fundamentals' to which Stapledon then referred his readers were, firstly, that the land was the source of being. It was their 'mother'. Thus, he opened *The Way of the Land* by writing that,

> The earliest and most vivid memories of my boyhood were long drives with my mother in the by-ways of North Devon. My mother had an unusually keen appreciation of the beauties of nature. It was the scene as a whole – a stupendous sunset, a starlit sky, a massive display of autumn tints, or an expansive view of sea and cliffs – that always most deeply affected my mother, for she was a woman who seemed intuitively to understand the ways of nature and she always saw the intricate pattern as one majestic whole. My mother's reactions to nature's changing moods and tenses were tinged with an almost forbidding sense of awe and it may truly be said of her that she humbly worshipped at the shrine of nature.
>
> I was devoted to my mother and grew to manhood deeply influenced by a mental intimacy with her, founded on sharing experiences of a character and to a degree unusual between mother and son. All my life I have been profoundly affected by natural beauty and if I have not acquired my mother's sense of awe I have always felt a strong sense of shame for all the crimes against beauty perpetrated by man.
>
> As I grew older and, as an agricultural scientist, began to study nature in operation, I was early forced to realize that to squander or abuse her riches was as great a crime as to offend against her unwritten laws of beauty.[22]

Stapledon translated this sense of inseparability from his nurturing mother into a universal, nurturing relationship between humanity and nature. As such, those farmers and agricultural scientists who regarded 'the plant merely as raw material, the soil as a factory, and the animal as a machine' were incapable of truly understanding the land and its productive potential.[23] This static understanding, which he viewed as complicit with patriarchy, capitalism, and imperialism, stifled all sense of growth, as well as all sense of mutual dependence and its obverse, individuality. He wrote that

> Where the land is concerned, what are the fundamentals, and with what materials are we to build? The fundamentals, I stand convinced, are not material, they are spiritual. When I say spiritual I mean little

> more than non-material, something that is not food, is not houses, is not roads or railway tracks, is not factories, but something different from that which most of us regard as the chief contribution of our land surface to the well-being of the nation. When I say spiritual I really mean biological: the vital, living, and essentially individual part of each and every man, the part which each of us shares with all living things, not our robot characteristics which are peculiar to man and which if not properly vitalized can only lead to degradation.[24]

Drawing on D. H. Lawrence and T. S. Eliot, he then argued that, 'man to be himself, to understand himself, must sometimes escape from man and mingle humbly, freely, gladly with other living things and with the universe – without thought, without inquiry'.[25] Stapledon was adamant, however, that his celebration of the natural world as the nurturing, if not motherly, principle of life and being, should not be understood as a nostalgic call for a return to some pristine, Arcadian way of life. He was not advocating a return to the womb. Instead, he sought to articulate an aesthetic reading of evolutionary theory and genetics. He wrote that 'nature is nothing if not artistically harmonious in her blendings of her numerous, conflicting and frequently savage purposes'.[26]

Stapledon's second, corollary, fundamental was that nature, and more specifically the land, was anything but fixed and unchanging. Humans and the land were so engaged in a process of destruction and recreation that focusing on a past in which there was a land without humans was completely misguided. What was important was instead greater attention to the dynamic interaction between humans and the land:

> You cannot preserve and progress . . . progress must always come before preservation . . . by progress, I expect I mean creation and creativeness, and in the widest biological sense; and that is in the sense of a universal stepping from achievement to achievement, but of achievements always framed against the background of harmony

> and beauty. The natural scene is always beautiful, but by natural scene we are apt to think of the scene unaffected by the doings of man: forgetting that man himself (ironically labelled *Homo sapiens*) is a biological phenomenon, and consequently there is no reason why the natural scene, let it be affected never so much by the doings of *Homo*, should not always be beautiful. Indeed I stand convinced that beauty is the hall-mark of creativeness. I mean beauty in all-embracing sense: in the sense that a hippopotamus may not appear to be a very decorative animal when seen in a zoo, but that, in his natural surroundings, his presence does nothing to render the scene less beautiful, nor does he himself appear in the least incongruous. Let it be granted that beauty is, in very fact, the hall-mark of creativeness, then if the so-called achievements of *Homo* are purchased at the price of a wholesale destruction of beauty, man is progressing not at all; he is, in fact, caught up in the clutches of the Devil, and the Devil only knows what his ultimate fate will be . . . Our mistake has been that we have not harnessed our new powers to the creation of beauty or in the service of love.[27]

Equally importantly, the converse idea that humans could enjoy any meaningful existence outside a relationship with the land was just as empty. Knowledge and awareness of one's own existence depended on both openness to the diversity and change that characterised the land and direct, unmediated experience of such diversity and change. Stapledon thus had no time for agricultural and medical scientists' belief in 'self determinism', that is to say, agricultural and medical scientists' belief that they could transcend variety and unpredictability by 'discovering' some immutable lawfulness of nature.[28] In his view, the fundamental tenets of modern biological science, namely repetition, statistical analysis and the 'evil genius of standardisation', were symptomatic of a closure that directly contradicted the fundamental nature of its object of study, life.[29] As he pointed out, 'we are beginning to study life: and what is life? The last thing that life is, is a cold average. There is no such thing as an average – no such thing as the man in the street. That is the spirit of modern biology: variation, difference, personal uniqueness.'[30] Echoing the conflicts between Sir Frederick Gowland Hopkins and Lord Moynihan, he then called into question where exactly true 'knowledge' was produced, asking whether 'Harley Street specialists, and for that matter country practitioners, contributed anything or not to the sum of scientific knowledge'.[31] His answer was that they most decidedly did, because their knowledge was grounded in 'experience'. He thus wrote that

> As always . . . the greatest and the final hope is the farmer himself, for he at least is untrammelled by the technique of science, and is not slave to the fashions current in science, while his major training is not in collecting data, but in the gentle art of unadulterated observation. Just because, therefore, of the immense accumulation of scientific knowledge, so much of it half digested in the practical sphere, never so urgently as at present has there been such a necessity for an abundance of well-informed, originally minded and affluent pioneers, men willing and eager to transgress against every canon of good husbandry, and to explore, and almost *de novo*, the whole field of rotation of crops, and the whole idea of rotation of pastures of different types and of stock over the surface of the farm.[32]

Stapledon then drew on contemporary research in animal behaviour to articulate further his definition of this more adequate attitude toward life and its attendant force of self-creation. He argued that

> Sympathy, sense, and feeling must necessarily be brought into play to assist to fill the gaps. Thus the man who feels his way into understanding as well as learns his way in is the only man who is competent to deal with the problems of life and therefore the problems of agriculture. It would indeed be a rash man who would deny, out of hand, the possibility of man having a sympathetic feeling for the plant world, and although we need not accept the etheric principle of the followers of Rudolf Steiner, if supremely honest with ourselves, I think we shall have to admit that there is much justice in some of the basal tenets held by and acted upon by the anthroposophists . . . to 'see things together' and to be 'able to reunite the things that belong to one another'.[33]

Thus, 'knowledge' was, at its best, a 'fiction' that facilitated 'the drive which imperatively demands activity'.[34] 'Wisdom', however, was the ability to recognise and live by this imperative, by constantly experimenting, ultimately to bring 'man, animal life and plant life into . . . one harmonious and purposeful activity'.[35]

Stapledon took his two fundamental tenets to heart by writing himself into his beloved Welsh highlands. As he put it,

> All I do in the hills is to add more tints of green to the rather limited range of that beautiful colour normal to hill land, and to render the general scene that much more complete by a not inartistic touch from the hand of man; nature merely gains more scope from the display of her infinite glories through my informed, albeit utilitarian, co-operation.[36]

In sum, for Stapledon, the land was not a space to be conquered by man, but place of perennial transformation where the truth of the human condition, variety, change and open orientation toward the future, was most evident. The land was the material manifestation of what Bergson called 'élan vital', and Stapledon renamed as the 'spirit of place'.[37]

This said, if Stapledon's celebration of life was strongly inflected by emphasis on action orientated toward the new, he also maintained that the critical importance of this understanding of being could only be appreciated by looking backward. Thus, Stapledon also wrote that,

> The views I have been bringing forward may seem to be those of unadulterated primitivism. I think otherwise; not for a moment would I argue that progress can be born alone of harping back. I do not merely desire to harp back, for it is one thing to declare that all is won if we will but look downward and backwards and quite another to assert, as I have been endeavouring to assert, that those who wish to see all they possibly can by looking forwards and upwards will obtain a better mental perspective if they will first glance backwards and downwards.[38]

In other words, Stapledon was caught ambiguously, if not contradictorily, between the past and the future.

Contemporary echoes

The Way of the Land is an ambiguous text. Of course, such ambiguity could be explained by focusing on the diversity of audiences to which Sir George Stapledon spoke over the twenty-three years during which the individual, constituting essays were written. This is exactly what I had done earlier, and is the reason why *The Way of the Land* had sat on my bookshelves uncut and unread. Yet, such an approach would have to ignore the very material way in which these essays were eventually bound into a single book, *The Way of the Land*, which forced and continues to force their reading as a single narrative. More importantly still, however ambiguous, if not contradictory, this highly popular text literally constituted 'Sir George Stapledon' as an important voice in contemporary arguments within the Council for Preservation of Rural England and the National Trust. Strikingly, Stapledon's position on proposals to establish national parks in the Pennines and the Lake District went to the heart of today's debates over these national parks' multiple and often contradictory functions as farmland, tourist amenities, heritage sites and ecological preserves.

Stapledon argued in *The Way of the Land* that the preservationist thrust of the National Trust had to be resisted. He wrote that

> If the Trust remain passive landowners of too large an area, by their very passiveness they will do harm, for in the meantime rural Britain is bleeding to death because nowhere on a sufficient scale will anybody do that essential 'something' that lies between doing precisely nothing and the wrong thing too thoroughly.[39]

Stapledon then explained what this 'something' was by noting that the Council for the Preservation of Rural England effectively negated the 'spirit of place', as it sought to preserve the 'way of the land' by forbidding any modification of what he viewed as outdated modes of agricultural production.[40] His counter-proposal was to allow instead a perpetual, 'enlightened' transformation of agricultural production throughout the national parks and other protected areas, so as to create a living landscape for tourists to remember their true nature. Today, as David Rose has noted in his assessment of the crisis over foot-and-mouth and the destruction it has wrought in the Lake District, we are still struggling with this issue. As the sheep disappear and we consider what its impact might be on the landscape, we realise that we have in fact chosen an arbitrary moment in the ever-changing relationship between humans and the world of things, and then transformed that moment into something immemorial and completely other, 'nature'. Nothing could be more contradictory of our relationship to the world of things.

Perhaps, then, we need to recognise contradiction not as exceptional, but intrinsic to being modern. Stapledon was at one with the grasses of the Welsh highlands. The land was Stapledon's counterpart to Francis Ponge's text, and, as such, Stapledon's own textual creation was bound to be inadequate to the totality he sought to capture. All that he could do, to be true to life, was to continue writing. Similarly, for the contemporary geneticist, the gene is paradoxically both the trace of a history that has determined who we are, and the very principle that allows him or her to believe that this history might be erased, and so allow him or her to start anew. As I noted earlier, Paul Rabinow advocates a new ethic of being that is more adequate to this supposedly unprecedented situation. This ethic must be grounded in a 'will to experiment' that is much more open to the future and the heterogeneous than is the ultimately humanist 'will to knowledge'. Being a witness to the unfolding of the age of genetic engineering is then a matter of being a more modest, 'disinterested' observer, a stance which Rabinow describes as

> An experimental mode of inquiry . . . where one confronts a problem whose answer is not known in advance rather than already having answers and then seeking a problem.[41]

The truth of *Homo* lies not in *sapere*, but in *experior*. Pursuing further the analogy between the geneticist and the historian, this then means that the archive should not be conceived as a place of recognition, but as a place of experiment in transformation. The historical actor is a fiction created in our encounter here and now with the infinite and multiple world of things. Remembering the dismembered can never return 'ownership of themselves' to the dispossessed, but constructs something quite different, a new historian.[42] Rejecting this fundamentally constructed nature of the historical actor, by imagining that to argue so is to do violence to the dispossessed, is a denial of life 'here and now', in the world of things. It is an act of misrecognition that bears the hallmark of nostalgia, fear of the future, if not the hallmark of death. I take it that this was the fuller meaning of Friedrich Nietzsche's statement:

> I trust in youth that has led me aright when it now compels me to protest at the historical education of modern man, and when I demand that man should above all learn to live, and should employ history only in the service of the life he has learned to live.[43]

Yet, the very location of the historiographical act of remembering in the 'here and now', also means that historical critique will then always be complicit with the 'here and now'.

The history of capital?

Thinking about the ambiguities of the preposition 'of' – a history of capital or capital's history? – it is clear that George Stapledon's critique of modern forms of knowledge and their complicity with patriarchy and capitalism was a celebration of the creative destruction inherent in what he called 'life'. As he put it, 'if we have to destroy, what matter, so long as we rebuild something of greater beauty than before, and something of greater spiritual and creative significance?'[44] Strikingly, however, these words resonate with Karl Marx and Friedrich Engels' famous passage from the 'Manifesto of the Communist Party':

> The bourgeoisie cannot exist without constantly revolutionising the instruments of production, and thereby the relations of production, and with them the whole relations of society. Conservation of the old modes of production in unaltered form, was, on the contrary, the first condition of existence for all earlier industrial classes. Constant revolutionising of production, uninterrupted disturbance of all social conditions, everlasting uncertainty, and agitation distinguish the bourgeois epoch from all earlier ones. All fixed, fast-frozen relations, with their train of ancient prejudices and opinions, are swept away, all new-formed ones become antiquated before they can ossify. All that is solid melts into air, all that is holy is profaned, and man is at last compelled to face with sober senses, his real conditions of life, and his relations with his kind.[45]

This suggests that Stapledon's celebration of creative destruction may have been complicit with the very spirit of capitalism that he sought to criticise. The similarity is not a matter of coincidence. As Gilles Deleuze and Félix Guattari, those other, more renowned readers of Henri Bergson, note, the emergence of the subject 'I' is the first step in a 'universal history' that leads inevitably to the emergence of capitalism. The subject 'I' comes into existence in relation to a multiplicity of other discrete entities, whose totality, given the nature of all these entities' foundation as arborescent extrusions, feeding on the rhizomatic energy of life, is always inadequate to the true nature of the world. The spontaneous desire for association, which lies at the heart of the primordial, rhizomatic flow, is then transformed into desire for something that the subject 'I' lacks. 'Desire' becomes 'need'. The 'transcendental signifier', the 'One', or 'God', then comes into being. However, once the 'transcendental signifier' is removed, once 'God is dead', self-realisation becomes a goal that can never be fully satisfied. Herein lies the dynamism and productivity of modern capitalism. Stapledon's invocation of 'life' partook of these negative dynamics of desire. His role in establishing the contemporary organisation of agricultural and food research that has brought us 'Frankenstein foods', and all that it entails for the 'future of the human', speaks to its inevitable outcome. Significantly, Paul Rabinow's account of the fate of knowledge and recommendation of a 'will to experiment' that is more open to the future is equally complicit.

In *French DNA*, Rabinow draws on Jacques Le Goff's *Your Money or Your Life* and *The Birth of Purgatory* to understand the profound anxiety that would seem to characterise French debates over the decoding of the human genome and its promises of redemption. He identifies the atmosphere of suspicion surrounding the alliance between capital and the decoding of human genome with a historical fear that relations of exchange will inevitably undermine the moral bonds between humans. This fear lies at the heart of the Christian condemnations of 'usury' that are the subject of *Your Money or Your Life*. Rabinow argues further that the consequent protracted debates over 'how best to bring capital, morality, and knowledge into a productive and ethical relationship' take the form of a 'purgatorial discourse'.[46] Purgatory, an invention of the mediaeval imagination, was the place of self-examination and purification for the less than virtuous, that is, the majority of Christians, before they could be admitted into Heaven. Similarly, the overriding presumption in the debates over the decoding of human genome is that, given time enough, it might be possible to arrive at a proper and moral response. As the Pope himself argued in *Evangelium Vitae*, more research is needed. Presumably, the full extension of the Papal argument is that humans might one day free the world of that evil that is illness and mortality. Yet, as Rabinow notes, a superabundance of time is exactly that which a properly secular and materialist human does not enjoy. Our first responsibility is to those around us, embodied people whose life is still all too finite. From this perspective, the endless debates over 'how best to bring capital, morality, and knowledge into a productive and ethical relationship' are fundamentally conservative. They foreclose the future, for example, by depriving the parents supporting the efforts of Association Française Contre les Myopathies promote research on the human genome of hope for a world in which their children will not die. In sum, Rabinow's point is that, despite all the concern about the 'future of the human', the 'future' is in fact sacrificed to a transcendental idea of the 'good' that ignores the fundamental truth of our existence, namely our finitude.

Rabinow's argument is undoubtedly appealing. 'God is dead'. Yet, following Éric Alliez's *Capital Times*, a genealogical reflection which aims to articulate an understanding of time that is adequate to Deleuze and Guattari's universal history, one can also read *Your Money or Your Life* rather differently than Rabinow does. One might argue that Christianity abhorred usury because it literally sought to bring the future into account. The same goes for speculation and interest, the secular heirs to usury, which then lends capitalism one of its chief characteristics, namely, the commodification of time. Thus, in 1998, Monsanto invested in the future, not in the present portfolio of Plant Breeding International. The value of this future portfolio was so calculable that Monsanto was prepared to pay £320 million, banking that, after 'discounting' the future, this would yield more than might be gained by, for example, putting the £320 million in government bonds. Thus, if

Rabinow wishes to ground his criticism of Michel Foucault's 'will to knowledge' in Deleuze's *Foucault*, if not Deleuze and Guattari's reflections on universal history, there is a cost to not thinking about the dynamics that underlie Deleuze and Guattari's historical narrative. Rabinow fails to notice the meaninglessness of his evocation of a 'will to experiment' that is more open to the future. In the 'age of genetic engineering', there is no future, if this is understood as a time of redemptive possibility. Of course, history has a way of getting in the way. In 1998, no one could have foreseen how the *Daily Mail* would capture the public's imagination by turning to Mary Shelley's *Frankenstein*, and how this would disrupt Monsanto's carefully crafted corporate strategy. This, however, may not be what Rabinow has in mind when he evokes the future, by calling for

> An experimental mode of inquiry . . . where one confronts a problem whose answer is not known in advance rather than already having answers and then seeking a problem.

In other words, there is no 'will to experiment' without 'will to knowledge'.

If no one can escape complicity with capital's history, however, it would then be disingenuous not to extend the same criticism to everything that I have written thus far about 'remembering in the age of genetic engineering'.

Notes

1 For an introduction to Tiago Moreira's work, see 'Translation, difference and ontological fluidity'.
2 Ponge, 'My creative method', pp. 12–13 (my translation).
3 Engledow, 'Rowland Harry Biffen', p. 23; and Pal and Mukherji, 'Sir Rowland Biffen', p. 85.
4 This section is based partly on Palladino, 'The empire, colonies and

lesser developed countries as mirror'.
5 Stapledon, *The Way of the Land*, p. 119.
6 *Ibid.*, pp. 135 and 138–9.
7 *Ibid.*, pp. 146–7.
8 *Ibid.*, p. 196.
9 *Ibid.*, p. 29.
10 *Ibid.*, p. 209.
11 *Ibid.*, p. 55.
12 *Ibid.*, p. 171.
13 *Ibid.*, p. 37.
14 *Ibid.*, p. 207.
15 *Ibid.*, p. 172.
16 *Ibid.*, p. 158.
17 Norman Simmonds, personal communication, 19 July 1989.
18 *The Way of the Land*, p. 62.
19 *Ibid.*, p. 62.
20 *Ibid.*, p. 154.
21 *Ibid.*, p. 14.
22 *Ibid.*, p. 5.
23 *Ibid.*, p. 223.
24 *Ibid.*, p. 62.
25 *Ibid.*, p. 64.
26 *Ibid.*, p. 89.
27 *Ibid.*, pp. 106–7.
28 *Ibid.*, p. 73.
29 *Ibid.*, p. 230.
30 *Ibid.*, p. 144.
31 *Ibid.*, p. 14.
32 *Ibid.*, p. 197.
33 *Ibid.*, p. 222.
34 *Ibid.*, p. 221.
35 *Ibid.*, p. 222.
36 *Ibid.*, p. 110.
37 *Ibid.*, p. 88.
38 *Ibid.*, p. 65.
39 *Ibid.*, p. 85.
40 *Ibid.*, p. 89.

41 Rabinow, *French DNA*, p. 174.
42 Fissell, *Patients, Power and the Poor*, p. 15.
43 Nietzsche, 'On the uses and disadvantages of history for life', p. 116.
44 Stapledon, *The Way of the Land*, p. 108.
45 Marx and Engels, 'Manifesto of the Communist Party', p. 338.
46 Rabinow, *French DNA*, p. 20.

Conclusion: This isn't it …

Tomorrow, the first rough draft of the human genetic code will be published – one of the epic achievements of contemporary science. We will know the gene sequences that determine our mental and physical behaviour. We will have the tools that in decades ahead will allow us to understand how much of what we do is predetermined and how much is of our own free will.

The bourgeoisie cannot exist without constantly revolutionising the instruments of production, and thereby the relations of production, and with them the whole relations of society. Conservation of the old modes of production in unaltered form, was, on the contrary, the first condition of existence for all earlier industrial classes. Constant revolutionising of production, uninterrupted disturbance of all social conditions, everlasting uncertainty, and agitation distinguish the bourgeois epoch from all earlier ones. All fixed, fast-frozen relations, with their train of ancient prejudices and opinions, are swept away, all new-formed ones become antiquated before they can ossify. All that is solid melts into air, all that is holy is profaned, and man is at last compelled to face with sober senses, his real conditions of life, and his relations with his kind.

It has not escaped our notice that the more we learn about the human genome, the more there is to explore.

On *zoē* and *bios*

The three textual fragments, with which *Plants, Patients and the Historian: (Re)membering in the Age of Genetic Engineering* began, have framed my effort to draw to a close nearly fifteen years' work on the history of agricultural and medical genetics, ironically, just as we enter the 'age of genetic engineering'.

As I have been trying to suggest throughout the narrative now beginning to draw to a close, I have been puzzled by the relationship between agricultural and medical genetics, even though, or perhaps because, Michel Foucault would have argued that they are nothing but distinct examples of the assimilating discourse of the Enlightenment. This discourse is pithily summarised in Foucault's *Discipline and Punish*, by the reproduction of the frontispiece to Nicolas Andry's *Orthopaedia*. Here, any difference between the straightening of misshapen trees and the physical and moral education of children is erased. More recently, Giorgio Agamben has extended Foucault's all too compelling analyses of such erasure still further, to argue ultimately that the 'camp', in which the human is deprived of all the rights that might distinguish him or her from animals, or plants, is the future.[1]

Agamben articulates this apocalyptic vision by noting firstly that, where we would use the term 'life', classical culture distinguished between two forms of existence, *zoē* and *bios*. *Zoē* referred to the form shared by humans and animals alike, a form concerned with material sustenance and reproduction. Agamben uses the phrase 'naked life' to capture this meaning more fully. *Bios* referred instead to the ethical and political form of existence that was oriented toward the realisation of the virtuous citizen of the *polis* and, as such, was peculiar to humans alone. In other words, *bios* was distinctively oriented towards an end that is not of the 'here and now'. Agamben then argues that Aristotle's famous definition of the human as a 'political animal' opened the way for the 'camp' to become the contemporary model of governance.[2] Agamben's reasoning is that, on Aristotle's definition, the 'law' is understood as the human institution that guarantees the separation of human and animal, and thus also guarantees the possibility of politics. This means, however, that the authority of the 'law' rests on the possibility of its suspension and the reversion of human existence to the 'state of nature'. To put the point somewhat less delicately, this means that sovereign power is in fact predicated on an understanding of the human as *zoē*. Today, this horrific truth of the 'law' is everywhere obvious, as the 'refugee' emphasises how much our notions of 'citizenship' rest not on inalienable human rights, but on the contingencies of birth, on the contingencies of biology.[3] Although Agamben does not address directly the political significance of the decoding of the human genome, the many promises of its applications in civic life and the increasing concerns about the emergence of a 'genetic underclass' would suggest that politics is indeed becoming a business of managing the human in its most denuded existence, a double strand of deoxyribonucleic acid. Agamben opposes all this by appealing to Plato's argument that the 'law', or the 'soul' and hence the 'law', is in fact prior to all human institutions.

While Agamben's vision and the argument that underlies it are in many ways quite compelling, I remain as unconvinced as I am by the frontispece to Andry's *Orthopaedia*. My reservations stem firstly from Agamben's effective characterisation of Aristotle's definition of the human as a 'political animal' as an extension of the Sophists' categorical opposition of 'nature' and 'law'.[4] Yet, if one were to attend to Aristotle's *History of Animals*, and not just to *The Politics* or *On the Soul*, as Agamben does, it would be clear that, if the human was, for Aristotle, a 'political animal', a number of other animals beside the human also were 'political'.[5] This is not an effect of the 'indistinction' that Agamben sees as a consequence of understanding the 'law' as a human institution, but of Aristotle's understanding of animals as also endowed with some 'potentiality'. The difference between Aristotle and Plato can then be understood as not being over the nature of the 'law' as a human institution or as prior to all human institutions, but over the location of the 'law'. At the cost of some anachronism, one might say that, for Aristotle, the 'law' is immanent rather than trancendent. It is on this understanding that I would almost say that Aristotle was, firstly and foremostly, a philosopher of 'life'. If this is to stretch readings beyond credibility, I would want to note at the very least that Aristotle's effort to oppose the platonism and its accompanying political utopianism, by explaining the origins of human institutions with reference to the world of things, was fraught with ambiguities.[6] From this perspective, it then seems to me that Agamben is trapped by the very same terms of debate that he seeks to criticise. To argue, as Agamben does, that *bios* and *zoē* are fundamentally different is complicit with the opposite argument that they are indistinguishable. 'Nature' is not a 'given' into which humanity is now being merged, but something whose meaning is as unclear as the 'human'. It seems to me therefore that it is an exaggeration to believe that the future of the human 'as such' is today at stake.

More constructively, I would return to the Foucault of *The Order of Things*, to argue secondly that both the categorical distinction between *zoē* and *bios*, and the attempt to account for the latter in terms of the former, is characteristic of a much more narrowly circumscribed modern discourse, wherein the transcendental signifier is effectively removed from political discourse, *logos* becomes immanent, and (*zoē*)nomia is consequently transformed into (*bios*)logy.[7] Significantly, capitalism also sprang from this more historically specific reconfiguration, and, as Michael Hardt and

Antonio Negri argue in *Empire*, a sociological rendition of Gilles Deleuze and Félix Guattari's philosophical reflections on the history of capital, this was not a matter of accident. I would also note that this more circumscribed understanding of the relationship between *zoē* and *bios* would still have to ignore how the hermeneutic imperative of pre-modern discourse continues none the less to inflect contemporary biological science. Contrary to Agamben, the arguments between Stephen Jay Gould and Richard Dawkins can be understood as speaking to a persistent and still lively concern with the meaning of 'life' as such.[8] In fact, the excessiveness of 'life', which Gould and Dawkins have been equally unable to capture fully, is the very condition of possibility for the vitality of 'biology'.

The problem, however, is how to give an account of the relationship between *bios* and *zoē* that is adequate to that 'excessiveness' of life denied by the frontispiece to Andry's *Orthopaedia*.

From representation to process and performance

While a number of geneticists argue that the human genome is the repository of human history, the contemporary excitement surrounding its decoding lies more with the prospect of redeeming humanity from its captivity to history. Thus, in last year's Reith Lecture, Tom Kirkwood argued that the length of human life is determined by the logic of 'selfish genes', which seeks to balance the costs of the genes' reproduction and the maintenance of our corrupt and disposable bodies long enough to ensure such reproduction. Consequently, it is only a matter of time before we can alter the balance in our favour and live in the eternal present conjured by the geneticist Stephen Jones, in a television advertisement for the personal insurers Equitable Life. This may all be highly speculative. In 1995, however, Lois Rogers wrote far less speculatively that

> British doctors will for the first time use a test to select cancer-free babies . . . Embryos of a woman with a high risk of passing on a form of bowel cancer will be screened and only healthy ones will be re-implanted. The same technique is likely to be used within two years to screen test tube embryos for a predisposition to inherited breast cancer.[9]

This much more limited promise of genetic engineering, which impelled my initial delving into the history of medical genetics, has now been realised, and we can look forward to still more radical applications. In the meantime, protesting voices are dismissed by a new generation of political leaders who call on the public to forget past attempts to transform the human genetic constitution. We must now enter the new world of genetic engineering without hesitation or trepidation, or we will be left behind in the advance on this newest human frontier. Cultural commentators as different as Donna Haraway, Dominique Lecourt, and Paul Rabinow add weight to this invitation. They do so by charging those who would object of advocating the most dangerous conservatism with their appeals to 'History' or 'Nature', if not 'God'. History is certainly losing the critical power it once held, as historians seek to become more politically relevant by projecting the present ever more explicitly into the past, thus constituting the past as just another space of representation. Nature, instead, is revealed increasingly as an artefact, whose creation as an immemorial space is threatened by its expansive commodification. Lastly, the human genome becomes the newest 'book of life', now fully decoded. As Richard Dawkins has put it, this achievement is best understood as the rendering of the 'word' into 'flesh'. The human is truly becoming its own maker and the measure of all things, but at a cost: it is no longer clear what it means to be human, and transgression is so pervasive as to become meaningless.

This uncertainty about the meaning of being human was heightened as I began to write about the relationship between *bios* and *zoē*, and noticed William Gibson's *Neuromancer* sitting on my bookshelves. In *Neuromancer*, the present alliance of genetic engineering, information technology, and multinational corporations is projected into an indeterminate future. It conjures a world where the material culture that sustains the boundaries between the past, the present, and the future is erased. If memory has a referent in this world to come, it is nothing but a discrete virtual space in a larger bio-informational matrix. It is no longer the 'archive', the principle of formation of both memory and political order, except in the most anodyne sense of the word, as the irreversibility of time gives way to the symmetry of spatial relationships.[10] All that there remains, is an eternal present, and, consequently, there no longer is any possibility of 'becoming'. Of course, the power of this image rests on a sense of living in a time when the past is still something altogether different, against which the future can be contrasted to dramatic rhetorical effect. On the other hand, by imagining Victorian England as a technocracy led by 'Lord Babbage' and 'Lord Darwin', Gibson and Bruce Sterling's *The Difference Engine*, which sat next to *Neuromancer*, recreates the political radicalism of nineteenth-century English society more engagingly than any more conventional historical accounts. This so engaging narrative, however, echoes *Neuromancer*, which is perhaps not surprising since Charles Darwin and Charles Babbage could legitimately be regarded as the 'fathers' of the 'age of genetic engineering'. *The Difference Engine* and *Neuromancer* thus raised difficult questions about the relationship between the then that was and the then that will be. Lastly, Orhan Pamuk's *The White Castle*, which also sat on the bookshelves, turns to the past, this time to imagine an encounter between Western technological genius and Oriental decadence, in the aftermath of the Battle of Lepanto, a fundamentally important moment in the history of western politics and metaphysics. Here, the meeting of the world of mind and the world of the body, each fascinated and revolted by the other, leads to a disturbing loss of centre, disturbing through the reader's interpolation of a question about the identity of the narrator. In sum, these literary texts seemed to speak directly to the problem of remembering the origins of our present situation, and thus to the possibility of giving an account of the relationship between *bios* and *zoē* that is adequate to the 'excessiveness' of life.

It seemed to me, however, that Michel De Certeau's recollection of Jean de Labadie, in the concluding chapter of *The Mystic Fable*, spoke directly to the challenge with which I was confronted.

> A man of the South . . . Labadie went north . . . From Guyenne, where he was born and became a Jesuit, he went to Paris, Amiens, Montauban, Orange . . . then thought perhaps he would go to London, no, it was Geneva, then the Netherlands, Utrecht, Middleburg, Amsterdam, then farther, to Altona in Denmark, where he died... The inner journey was transformed into a geographical one. Labadie's story is that of indefinite space created by the impossibility of place. The stages of the journey are marked by the 'religions' he passes through, one by one: Jesuit, Jansenist, Calvinist, Pietist, Chiliast or Millenarian, and finally 'Labadist' – a mortal stage. He passes on. He cannot stop.[11]

For Labadie, the truth of 'becoming' was affirmed in the ever more manifest impossibility of comprehending its end. Yet, De Certeau, toying with historiographical conventions, by inserting references to contemporary plays such as Jean-Paul Sartre's *The Condemned of Altona* where the historian might have expected references to documents from the seventeenth century, also casts doubt on the historical location of Labadie. Furthermore, he reminds his audience that there can be no 'self' without 'others' by writing that

> [Labadie] brings to mind John the Baptist, the walker, sculpted by Donatello, in the instant at which movement is loss of equilibrium. That falling becomes walking if it happens to be the case that a

second place exists to follow the first, but the artist, by isolating the figure, makes that hypothesis uncertain. How can we be sure whether he is falling or walking? . . . To walk is to propel oneself outward, to jump out of the window. Labadie falls out of the places that cannot hold him. ... It ends up being a story because each time, miraculously, other places 'receive' him, or, more precisely, as he constantly repeats, they 'preserve' his body from falling by 'supporting' him.[12]

Labadie's wandering and its conditions of possibility thus became crucially important for those like myself, those who would seek to understand the making of our present situation and all that it would appear to entail for the 'future of the human'. I will try to clarify Labadie's answer by imitating Babbage's 'difference engine', which sought to approximate the 'real' by iteration and addition.

Biology, history and melancholia

Let me begin anew, with two fragments from Walter Benjamin's writings on philosophy and history.

> A Klee painting named 'Angelus Novus' shows an angel as though he is about to move away from something he is fixedly contemplating. His eyes are staring, his mouth is open, his wings are spread. This is how one pictures the angel of history. His face is turned toward the past. Where we perceive a chain of events, he sees one single catastrophe which keeps piling wreckage upon wreckage and hurls it in front of his feet. The angel would like to stay, awaken the dead, and make whole what has been smashed. But a storm is blowing from Paradise; it has got caught in his wings with such violence that the angel can no longer close them. This storm irresistibly propels him into the future to which his back is turned, while the pile of debris before him grows skyward. The storm is what we call progress.[13]

> A philosophy that does not take into account the power to prophesies from coffee grains cannot be true philosophy.[14]

When I first read the first fragment, I was mesmerised by the evocation of movement and desire. When I finally found a reproduction of Paul Klee's painting, however, I also happened to be watching some very early cinematographic experiments by the Lumière and Pathé brothers – 'Leaving Jerusalem by train' and 'Dream and reality'. Significantly, these efforts coincided with the rediscovery of the mendelian principles of inheritance. I came to see the angel of history as ensnared between two genealogies. On the one hand there was the modern dedication to capturing the experience of movement, of trains moving ever faster past bicycles and horse-drawn buggies, in a way that painting or photography never could. On the other hand there was the modernist dedication to shattering the difference between the real and the imaginary. This is perhaps best consummated in Larry and Andy Wachowski's *The Matrix*: in the exchange between Agent Smith and Cypher, meat, historically the signifier of the real, no longer secures the boundaries between the real and the imaginary or the virtual. The tension between these two genealogies was central to what I now appreciate as my impossible effort to (re)member the materially dismembered.

Some years ago, spurred by Stephen Jay Gould and Richard Lewontin's politically charged critiques of genetic determinism and celebrations of historical contingency, I began to work on the history of genetics by turning my attention to the Road Improvement and Development Fund Act. This largely forgotten piece of legislation, enacted in 1909, established and did much to foster the growth of institutions such as the Plant Breeding Institute, the Welsh Plant Breeding Station, and the Scottish Plant Breeding Station into preeminent centres for genetic research. Its remarkable support for the then novel science of genetics was impelled by a vision best captured by William Bateson, who first coined the word 'genetics'. He recommended genetics to the Board of Agriculture because

> There is something that will come out of [genetics] that will equal, if not exceed . . . anything that any other branch of science has ever discovered . . . A precise knowledge of the laws of heredity will give man a power over his future that no other science has ever endowed him.[15]

I sought to inject a note of scepticism by returning John Percival and Edwin Sloper Beaven to history. Both were involved in this institutional organisation of genetic research, and yet they were deeply sceptical about the promise of genetics. Even as late as in 1922, Beaven was writing to Percival:

> There is a F4 family of 20000 plants – not one (I believe) homozygous. How is that for orthodox genetics? Not much I think. Bateson is coming to see it soon but I wish you could come first to give me a few tips with which to comfort him.[16]

Percival and Beaven's shared scepticism was grounded in a historical understanding of plants' adaptation to their local environment. This historical understanding, which established the conditions of possibility for the emergence of genetics, was broken in part by erasing Beaven and Percival from the historical record. Sifting through those 'coffee grains' that are the innumerable and scattered archives I visited in my travels between the four cardinal points marked by Dundee, Warminster, Aberystwyth and Norwich, I sought to expose how their dismembering was the inevitable product of a momentous historical process. This process explains why, in 1998, Monsanto paid £320 million to acquire Plant Breeding International, the heir to the Plant Breeding Institute. Echoing Gilles Deleuze and Félix Guattari, as well as Thomas Hughes, if the reporters for the *Financial Times* were 'surprised' by this sum, it was simply because they forgot that the gene was a reterritorialising 'network' technology, intimately linked with the history of capital, from the day it was born.[17]

Yet, if history is as important as Beaven and Percival once suggested, and I, as a historian, would like to believe, remembering Beaven and Percival is a strange affair. As Benjamin himself understood all too well, restitution is a matter for the Messiah alone. In an age that supposedly no longer believes in the messianic, remembering can then only be a melancholic affair.

Bio-power, biography and the return of the subject

Picture a historian not particularly interested in biography being asked by a friend why they often turned to biography in their various attempts to understand the making of the 'age of genetic engineering'. Allow me to play on the ambiguities of the pronoun 'they' as I write that this historian begins to remember from whence they came: there is no 'self' without 'others'.

The natural sciences are virtually synonymous with modernity, for no other area of human endeavour has been more thoroughly dedicated to demystifying the world. For the past thirty years, historians of science have lived in imitation of their preferred object of inquiry, as they relocated historical agency from uniquely insightful individuals onto social institutions, and then from social institutions onto the material practices of scientific inquiry. Echoing the logic of genetics, which recognises no boundaries between plants and humans, this historian moved their attention from the arguments between Sir Rowland Biffen, John Percival and Edwin Sloper Beaven, over the significance of genetics for plant breeding, to the arguments between Percy Lockhart Mummery and Georgiana Bonser. In 1935, Lockhart Mummery, a surgeon at St Mark's Hospital, in London, argued that

> Genetics must inevitably become the most important social and scientific problem of the next few decades, if the human race is to make any serious progress towards something better.[18]

At the same time, however, he also called into question the use of inbred mice, one of the mainstays of research in medical genetics, by writing that

> Experimental results cannot be applied too closely to the problem in man, because the conditions of mating necessary to demonstrate [genetic influence] in mice never obtain in any civilised community of mankind.[19]

Bonser, a researcher in the Department of Experimental Pathology and Cancer Research at the University of Leeds, replied to this criticism in a way that recalls the frontispiece to *Orthopaedia* and the poster enjoining Bonser's contemporaries to sow only 'healthy seeds', by simply stating that

> No one would deny that the mendelian laws are as applicable to the human as to the tall and short peas which Mendel used in his original experiments . . . Similarly, in the study of breast cancer the use of inbred mice is an invaluable aid to the elucidation of the problem in man.[20]

By focusing on Lockhart Mummery's material practices around St Mark's Hospital, however, this historian argued that Lockhart Mummery and Bonser were equally involved in the objectifying project of medical genetics. In fact, Lockhart Mummery's practices paved the way for the identification of the now fully sequenced mutation of the APC locus that lay at the centre of *The Sunday Times* report of the first pre-implantation embryo screening. This identification entailed the literal entrapment of patients and their relatives, an entrapment epitomised in a letter written by Richard Bussey, the Director of the Polyposis Registry at St Mark's Hospital, to a corresponding general practitioner:

> We have sent our beaters out after some polyposis children who have not been seen for a while or not at all. One of these patients has apparently been caught in your net.[21]

These lines, written as recently as in 1977, also speak, however, of an incessant, but constantly self-defeating, effort to deny humans their political existence. This persistent discursive misfire was none the less productive insofar as it resulted in the introduction of critical genetic concepts such as 'penetrance' and 'linkage'. This 'return of the repressed' also produced the return of an authorial voice that could no longer fulfil the historiographical and sociological demands of being what Donna Haraway calls a 'modest witness'. 'I' concluded that, notwithstanding the widely shared use of linguistic terms such as 'gene' and 'family history', the corresponding practices of medical practitioners and patients at St Mark's Hospital were, and are, so diverse and heterogeneous as to defy any Foucauldian notion of a 'discourse of the gene'. To suggest, therefore, that a locally contingent use of these terms amounts to absolute complicity with a discourse that would appear to be opening the door to a new eugenics, this time allied to the logic of advanced consumer capitalism rather than the corporatist state, would itself be complicit with this logic. In other words, the 'will to narrative' perpetuates the very violence that remembering is supposed to exorcise.

As I began to articulate these thoughts about remembrance, there was a simmering disquiet about the sociological approach to the history of science. Arguably, Adrian Desmond and James Moore's widely acclaimed biography of Charles Darwin exemplified this approach.

Darwin's name is indelibly associated with a return to the *zoē* of human existence. This association is not just a matter of metonymy, but is the symptom of persistent attachment to genius, to the remarkable ability of some individuals to withstand the destructive wind of progress. Desmond and Moore sought to demystify such genius by reducing Darwin and his ideas to nothing more than a product of Victorian society and values. The consequent disquiet erupted furiously in the wake of Gerald Geison's biography of Louis Pasteur, *The Private Science of Louis Pasteur*. Since then, a number of historians, and Thomas Söderquist in particular, have highlighted the self-defeating nihilism of this historiographical approach. Significantly, Thomas's reading of *Darwin* as an example of the limitations of the sociological approach to the history of science amply illustrates the indeterminacy of texts: Robert Young has criticised Desmond and Moore for being insufficiently attentive to the sociology of knowledge. Reading speaks volumes about the reader. Thomas called for a new hermeneutics, grounded in biographies focusing on historical actors' existential struggles. He argued that this would return historical narrative to the edifying function it played before it became a human science, now lacking any reason or purpose. I was not immune to this call and turned anew to Lockhart Mummery. I now took my cue from the frontispiece to Lockhart Mummery's *Nothing New Under the Sun*. As I noted earlier, in this frontispiece, a one-eyed bird and fish, one-eyed perhaps to symbolise the modern desire for omniscience, are superimposed on an aeroplane. This composite figure seems to have flown too close to the sun, and, like Icarus, is falling into the stormy sea below. I then proceeded to note that Lockhart Mummery was clearly committed to the project of medical genetics, and that this was linked inextricably to a eugenic vision of the future of humanity. Yet, Lockhart Mummery's uneasy use of metaphors, linking *bios* and *zoē*, also betrayed how humans remained nonetheless fundamentally different to the trees conjured by the frontispiece to Nicolas Andry's *Orthopaedia*. This seemed to me to offer, paradoxically, a way of returning to narrative the fractures and cacophony of St Mark's Hospital, by assimilating all actors into Lockhart Mummery's ambivalence, as they all confronted the consequences of the modern ambition to become the measure of all things. More importantly, however, I also noted how this biographical turn speaks loudly about the sources of the more general insistence on narrative closure.

The experimental life

Biography presupposes a transcendental subject, which is something so mysterious and alluring that Walter Benjamin once described it as a 'most fearful drug . . . which we take in solitude'.[22] The analogy is very useful, for we can then begin to understand how the contemporary return to the subject is complicit with the therapeutic culture of late capitalism. We live in an age when all anchors of the modern subject are no longer. In a play of words on Benjamin's essay on 'Art in the age of mechanical reproduction', Fred Botting calls it an 'age of techno-moral consumption'.

In 'The art of smoking', Botting focuses on the contemporary, paradoxical obsession with the elimination of smoking. As he points out, smoking could be considered the ultimate act of consumption, and thus a society that is as utterly dedicated to consumption as that of today should view smoking as the most legitimate of acts. The point is neatly captured by Oscar Wilde's aphorism: 'A cigarette is the prefect type of a pleasure. It is exquisite, and it leaves one unsatisfied. What more can one want?'[23] In articulating the dynamics of this paradoxical obsession with smoking, Botting then delineates a new discursive formation in which the production of perpetually desiring machines, ignorant of any transcendental meaning beyond the fulfilment of their earthly desires, requires an unprecedented intensification of moral surveillance. The surveillance of smoking is especially important because smoking betrays the fundamental emptiness at the heart of this discursive formation, which Wilde so neatly encapsulated. Significantly, such surveillance is mediated by technologies of the self that are themselves eminently consumable. Thus, we can respond to our alienation by reading a biography, which today is such a popular literary genre that in Britain alone over three thousand are published each year.

Biography is the salve for our longing to feel how it feels to be unique, and memorable. However, like the therapeutic drug that soon becomes an addictive drug, one dose of this salve can never be enough and soon leads to destruction. In seeking to capture the uniqueness of particular figures, that unparalleled ability to withstand the destructive wind of progress, that unparalleled ability to never be dismembered, by reducing these figures to the commensurate and mundane, the biographical enterprise demystifies exactly that which we desire. Thus, capturing the authentic, hermeneutically closed life is impossible, but the effort is none the less economically productive. We want more biographies still, to feel how it feels to feel. As Sarah Franklin has noted recently, the same goes for the gene as well. We want to know who we are, and the gene promises to tell us once and for all time. Thus the report by the International Human Genome Sequencing Consortium announcing the 'Initial sequencing and analysis of the human genome' closes with the following lines from T. S. Eliot's 'Little Gidding':

> We shall not cease from exploration.
> And the end of all our exploring
> Will be to arrive where we started,
> And know the place for the first time.[24]

Like Benjamin, I do not think that we can rise above the discourse of alienated subjects that compels the historian's impossible desire to 'stay, awaken the dead, and make whole what has been smashed'. I wish, however, to resist his politics of despair. As Franco Rella has noted, in his more optimistic moments, Benjamin called for a 'Copernican revolution' in the understanding of historiography, but we can no longer share his faith in that directed movement which is the stuff of trains, films and, of course, messianic historical materialism. We can, however, push for a new 'Copernican revolution' in which historiography is refashioned in a way that is more adequate to what Paul Virilio has described, in *Open Sky*, as our world of terminal velocity and digital oscillation.

Megan Boler has argued that ethical relationships, such as that sought by Thomas Söderquist, and all those who would advocate a return to Jules Michelet's 'resurrectionism', cannot rest on 'empathy', as this can easily degenerate into 'substitution'. We could then begin to reinstate the moral function of historiography to which Söderquist aspires, without, however, falling into 'substitution', by breaking the boundaries between actor and narrator, trading a life for a life, virtually. Yet, this too would entail an identification that is inconsistent with an anthropological turn, which, by drawing attention to material practices of historiography, would call into question the status of the lost monad that I was constructing out of disparate archival fragments and called 'Percy Lockhart Mummery'. All there is, are pieces of paper and photographs that are as silent as that most denuded form of life that is a double helix of deoxyribonucleic acid. Significantly, Francis Ponge neatly encapsulated the situation by writing that

> The variety of things is really that which constructs me. This is what I mean: their variety makes me, allows me to exist within the silence itself. Like the place around which they exist. But in relationship to one thing alone, having regard for each of them in their particulars, if I consider one of them alone, it annihilates me. And, if it is my pretext, my reason for being, if it is responsible for, and the root cause of my existence, this is, this can only be, thanks to a certain creation on my part regarding this thing. What creation? The text.[25]

Sir George Stapledon, the director of the Welsh Plant Breeding Station and architect of the technocratic refashioning of British agriculture that paved the way for 'Frankenstein foods', the phrase coined by the *Daily Mail* that eventually encapsulated public fears of genetically modified foods, lived exactly this life. He sought to constantly experiment, to bring 'man, animal life and plant life into . . . one harmonious and purposeful activity'. As he put it, with his genetic modification of grasses,

> All I do in the hills is to add more tints of green to the rather limited range of that beautiful colour normal to hill land, and to render the general scene that much more complete by a not inartistic touch from the hand of man; nature merely gains more scope from the display of her infinite glories through my informed, albeit utilitarian, co-operation.[26]

Nature, for Stapledon, was not a space to be conquered by man, but the place where the truth of the human condition was revealed. Stapledon's highlands were Ponge's text. Similarly, the single archival fragment, such as the frontispiece from *Nothing New Under the Sun*, is the site of origin of both the historical subject that was, and the historian that will have been. Only a narrative that instantiates itself in the endless oscillation between these single archival fragments, or Mummery and Stapledon's genes, and the future that will have been can participate in the drama of the 'here and now'.

This understanding of the archive and history, of the gene and life, brings me back to the decoding of the human genome. As Paul Rabinow has argued in *French DNA*, we can reject the hermeneutic imperative, by constructing and reconstructing ourselves through endless self-experimentation. He calls for:

> An experimental mode of inquiry . . . where one confronts a problem whose answer is not known in advance rather than already having answers and then seeking a problem.[27]

What the consequences of this 'will to experiment' will be, is not, and has never been, for us to decide, but for the future, when all these experiments will have been.

This isn't it . . .

> A spectre is haunting western academia ... the spectre of the cartesian subject. All academic powers have entered into a holy alliance to exorcise this spectre.[28]

Like the mythical Jean de Labadie, however, I, this strangely irrepressible authorial voice, cannot help but say once again 'this isn't it . . . !'

Paul Rabinow's invocation of a 'will to experiment', like Sir George Stapledon's assertion of self by immersion into the genetically modified hills of Wales, entails an impossible abdication of self. As Paul De Man wrote some years ago,

> The power of memory does not reside in its capacity to resurrect a situation or a feeling that actually existed, but is a constitutive act of the mind bound to its own present and oriented toward the future of its own elaboration.[29]

The 'will to experiment' is oriented inevitably to the realisation of something different, of something that is not of the 'here and now'. Žižek's *The Ticklish Subject*, in which Žižek provocatively rephrases the 'Manifesto of the Communist Party' and argues for the inescapable presence of the Absolute, has much to say about the refusal to acknowledge such orientation.

As Howard Caygill has noted in his essay 'Notes toward a metaphysics of the gene', the effort to decode the 'book of life' is nothing less than an effort to realise transcendence. Richard Dawkins amply confirms Caygill's analysis, as he celebrates the decoding of the human genome by writing 'We can all be proud of our species as it closes in on this summit of self-knowledge'.[30] Yet, with the increasing accessibility of genetic technology, and the hope that it offers that we might know – 'for the first time' – who we are, we will each become, in all our differentiating particularities, the measure of all things, and hence the measure of nothing. The 'we' evoked by the International Human Genome Sequencing Consortium is the 'empty centre' evoked by Oscar Wilde. As Žižek would note, such emptiness does, however, feed into the machine that promises to satisfy our every earthly desire, but never can do so – 'the more we learn about the human genome, the more there is to explore'. In fact, this machine must never satisfy our every earthly desire, but always defer it onto something else. It it were otherwise, the machine too would die. The name of that machine is 'capital', and Rabinow's invocation of a 'will to experiment' is absolutely complicit with its working. There can be no 'will to experiment' without 'will to knowledge'.

There is an alternative to this situation and all that it entails politically, and it seems to me that it lies with Labadie, rather than with the all too historicisable dogmatism of Žižek's St Paul or Lenin.[31]

Labadie's incessant wandering, marked by the *staccato* of his sequential, but not cumulative, refusals of the 'Jesuit, Jansenist, Calvinist, Pietist, [and] Chiliast or Millenarian', marks a fundamental incommensurability. This is the incommensurability between an ever expanding wilderness that will not allow Labadie, the literal and symbolic embodiment of *zoē*, to 'stay, awaken the dead, and make whole what has been smashed', and thus realise the *bios* of the *polis* to come. His increasingly lonely wandering then speaks to something excessive within *zoē* itself. Michel De Certeau commemorates this excess by closing *The Mystic Fable* with the following, rhythmic lines by Catherine Pozzi.

Très haut amour; s'il se peut que je meure
Sans avoir su d'où je vous possédais
En quel soleil était votre demeure
En quel passé votre temps, en quelle heure
Je vous aimais

Très haut amour qui passez la mémoire
Feu sans foyer don't j'ai fait tout mon jour,
En quel destin vous traciez mon histoire,
En quel sommeil se voyait votre gloire,
O mon séjour . . .

Quand je serai pour moi-même perdue
Et divisée à l'abîme infini,
Infiniment, quand je serai rompue,
Quand le présent don't je suis revêtue
Aura trahi,

Par l'univers en mille corps brisée,
De mille instants non rassemblés encor,
De cendre aux cieux jusqu'au néant vannée
Vous referez pour moi une étrange année
Un seul trésor

Vous referez mon nom et mon image
De mille corps emportés par le jour,
Vive unité sans nom et sans visage,
Coueur de l'esprit, ô centre du mirage
Très haut amour.

Most high love, if I should die
Without having learned whence I possessed you,
In what sun was your abode
Or in what past your time, at what hour
I loved you

Most high love that passes memory
Fire no hearth holds that was all my day,
In what destiny you traced my story,
In what slumber your glory was beheld,
Oh my abode . . .

When I am lost to myself,
Divided into the chasm of infinity,
Infinitely when I am broken,
When the present presently enrobing me
Has betrayed,

Through the universe in a thousand bodies shattered,
Of a thousand not yet gathered instants,
Of winnowed ashes windblown to the heavens' void,
You will remake for a strange year
One sole treasure

You will remake my name and image
Of a thousand bodies borne by days away,
Live unity with neither name or face,
Spirit's heart, of centre of mirage
Most high love

Catherine Pozzi and De Certeau's elegaical understanding of the relationship between being and the world reappears in Marilyn Frye's philosophical essay on 'To be and be seen'. Almost as if summarising the thrust of *Plants, Patients and the Historian*, she writes:

> This inquiry, about what is not encompassed by a conceptual scheme, presents problems which arise because the scheme in question is, at

> least in the large, the inquirer's own scheme. The resources for the inquiry are, in the main, drawn from the very scheme whose limits we are already looking beyond in order to conceive the project. This undertaking therefore engages me in a sort of flirtation with meaninglessness – dancing about a region of cognitive gaps and negative semantic spaces, kept aloft only by the rhythm and momentum of my own motion, trying to plumb abysses which are generally agreed not to exist and to map the tensions which create them. The danger is to fall into incoherence. But conceptual schemes have saving complexities such that their structures and substructures imitate and reflect each other and one can thus locate holes and gaps indirectly which cannot, in the nature of the thing, be directly named.[32]

The rhythm of all these lines speaks to the presence of the Other, the One, within *zoē* itself.

In sum, life is that 'constitutive absence' which the practices of 'biology', 'bio-power' and 'biography' seek to represent, by first distinguishing *bios* and *zoē*, and then reducing the former to the latter, but, at the same time, eludes 'biology', 'bio-power' and 'biography'. Life is that movement impelled by the tension between, on the one hand, the faith to which the lines from 'Little Gidding' speak:

> We shall not cease from exploration.
> And the end of all our exploring
> Will be to arrive where we started,
> And know the place for the first time.

On the other hand is the awareness that 'the more we learn about the human genome, the more there is to explore'. Drawing on Lily Kay's last thoughts about the writing of the new 'book of life', and echoing Roy, the all too human 'replicant' of *Bladerunner*, I will then begin to write my last words by wondering whether this new 'book of life' is not just another attempt to make visible this incommensurability, and whether it too will eventually 'be lost in time, like tears in rain . . .' Of course, as Lily would have noted, this latest attempt is not without its political consequences. It is then worth recalling how the One has recently made its presence felt among the protesters who have put themselves in the line of fire in Seattle, Göteborg and Genova, to protest against the myriad of small injustices, such as the production of 'designer babies' and 'Frankenstein foods', which taken together go by the name of 'globalisation'. Strikingly, these protesters have no coherent and well-defined political platform. All that they can say in unison is that 'this isn't it!' As such, sadly and perhaps tellingly, they have no place in Michael Hardt and Antonio Negri's reflections on the future of political opposition to the process of globalisation. I would claim, however, that their power lies in exactly this frustrating utterance. It is a protest against this world, fully engaged in this world, but orientated toward something not of this world. They desire (no)thing, but justice. They cannot then be reintegrated into political discourse. For once, that smoothly running machine that is capital does not know what to do, other than step up the violence. This, however, is another story, yet to be written . . . In the meantime, 'I' will close by saying 'thank you, Labadie'.

Notes

1 Agamben, *Means Without End*, pp. 37–45.
2 Aristotle, *The Politics*, p. 59.
3 Agamben, *Means Without End*, pp. 15–26.
4 Agamben, *Homo Sacer*, pp. 30–5.
5 Aristotle, *Historia Animalium*, pp 13–15; for Agamben's readings of *On the Soul*, see Agamben, *Potentialities*, pp. 230–2.
6 See Mulgan, 'Aristotle's doctrine that man is a political animal'; and Schofield, *Saving the City*, pp. 100–14.
7 See Cavanaugh, 'A fire strong enough to consume the house'; and Foucault, *The Order of Things*, pp. 250–302.
8 See Agamben, *Means Without End*, pp. 7–8.

9 Rogers, 'Doctors to create cancer-free babies', p. 24.
10 See Derrida, *Archive Fever*, pp. 1–5; and compare with Lynch, 'Archives in formation'.
11 De Certeau, *The Mystic Fable*, p. 271.
12 *Ibid.*, p. 272.
13 Benjamin, 'Theses on the philosophy of history', p. 249.
14 Benjamin, as quoted in Scholem, *Walter Benjamin*, p. 77 (my translation).
15 Bateson, 'Toast of the Board of Agriculture, Horticulture and Fisheries', p. 76.
16 University of Reading Archives (Reading): Percival Papers: Edward Sloper Beaven to John Percival, 20 June 1922.
17 Tait and Urry, 'Monsanto pays £320m for UK crop breeding business', p. 33.
18 Lockhart Mummery, 'Medical science and social progress', p. 1022.
19 Lockhart Mummery, 'Summary', p. 17.
20 Bonser, 'Influence of heredity on breast cancer', p. 456.
21 St Mark's Hospital (London): Polyposis Registry: Family 30: H. J. R. Bussey to M. Orr, 23 February 1977.
22 Benjamin, 'Der Sürrealismus', p. 213 (my translation).
23 Wilde, as quoted in Botting, 'The art of smoking in an age of techno-moral consumption', p. 78.
24 Eliot, as quoted in Lander *et al.*, 'Initial sequencing and analysis of the human genome', p. 915.
25 Ponge, 'My creative method', pp. 12–13 (my translation).
26 Stapledon, *The Way of the Land*, p. 110.
27 Rabinow, *French DNA*, p. 174; compare this with Agamben, *Means Without End*, p. 7.
28 Žižek, *The Ticklish Subject*, p. 1.
29 De Man, *Blindness and Insight*, p. 92.
30 Dawkins, 'The word made flesh', p. 11.
31 See Žižek, *The Fragile Absolute.*
32 Frye, *Politics of Reality*, p. 154.

Bibliography

Anonymous, 'A dream of AD 2456', *The Times Literary Supplement*, 15 February 1937, p. 102.
—— 'E. S. Beaven', *Times*, 20 November 1941, p. 7.
—— 'Seeds of dogma', *New Scientist*, 13 August 1987, p. 20.
—— 'The book of life: Gene science spells out our destiny', *Observer*, 25 June 2000, p. 28.

Abbott, A., *The System of Professions: An Essay on the Division of Expert Labor* (Chicago: University of Chicago Press, 1988).
Agamben, G., *Homo Sacer: Sovereign Power and Bare Life* (Stanford: Stanford University Press, 1998).
—— *Potentialities: Collected Essays in Philosophy* (Stanford: Stanford University Press, 1999).
—— *Means Without End: Notes on Politics* (Minneapolis: University of Minnesota Press, 2000).
—— *L'Ouvert: De l'Homme et de l'Animal* (Paris: Rivages, 2002).
Allen, G. E., 'Chevaux de course et chevaux de traite: Métaphores et analogies agricoles dans l'eugénisme américain, 1910–1940', in J.-L. Fischer and W. H. Schneider (eds), *Histoire de la Génétique: Practiques, Techniques et Théories* (Paris: ARPEM, 1990), pp. 83–98.
Alliez, É., *Capital Times: Tales from the Conquest of Time* (Minneapolis: University of Minnesota Press, 1996).
Althusser, L., *For Marx* (London: Allen Lane, 1969).
Arendt, H. *The Human Condition* (Chicago: University of Chicago Press, 1958).
—— (ed.), *Illuminations* (London: Fontana, 1973).

Aristotle, *Historia Animalium* (Cambridge, Mass: Harvard University Press, 1965).
—— *The Politics* (London: Penguin, 1992).
Austoker, J., *A History of the Imperial Cancer Research Fund, 1902–1986* (Oxford: Oxford University Press, 1988).
Barthes, R., 'The death of the author' [1968], in *Image Music Text* (London: Fontana, 1977), pp. 142–8.
—— *Camera Lucida* (London: Verso, 1993).
Bateson, W., 'Toast of the Board of Agriculture, Horticulture and Fisheries', *Report of the Third International Conference on Genetics* (London: Royal Horticultural Society, 1907).
Bauman, Z., *Modernity and the Holocaust* (Cambridge: Polity Press, 1989).
—— *Modernity and Ambivalence* (Cambridge: Polity Press, 1991).
Beaven, E. S., *Barley: Fifty Years of Observations* (London: Duckworth, 1947).
Bell, G. D. H., 'The *Journal of Agricultural Science*, 1905–1980: A historical record', *Journal of Agricultural Science*, 94 (1980): 1–30.
—— 'Frank Leonard Engledow, 1890–1985', *Biographical Memoirs of the Fellows of the Royal Society*, 32 (1986): 189–219.
Benjamin, W., 'Der Sürrealismus: Die Letzte Momentaufnahme der europäischen Intelligenz' [1929], in *Angelus Novus : Ausgewählte Schriften* (Frankfurt: Suhrkamp, 1966), Vol. 2, pp. 200–15.
—— 'Unpacking my library: A talk about book collecting' [1931], in Arendt, *Illuminations*, pp. 61–9.
_____ 'The work of art in the age of mechanical reproduction' [1936], in Arendt, *Illuminations*, pp. 211–44.
—— 'Theses on the philosophy of history' [1940], in Arendt, *Illuminations*, pp. 245–55.
Biffen, R.H., 'Modern wheats', *Journal of the Farmers' Club*, (1924): 1–18.
Bloor, D., *Knowledge and Social Imagery [2nd ed.]* (Chicago: University of Chicago Press, 1991).
Boler, M., 'The risks of empathy: Interrogating multiculturalism's gaze', *Cultural Studies*, 11 (1997): 253–73.
Bonser, G. M., 'The value of inbred mice in relation to the general study of mammary cancer', *British Medical Journal*, (1940), Vol. 1: 125–6.
—— 'Influence of heredity on breast cancer', *British Medical Journal* (1941), Vol. 1: 456.

—— 'Recent trends in cancer research', *Journal of the Medical Women's Federation*, 3 (1950): 22–5.
—— D. B. Clayson, J. W. Jull and L. N. Pyrah, 'The carcinogenic properties of 2-amino-1-naphthol hydrochloride and its parent amine 2-naphthylamine', *British Journal of Cancer*, 6 (1952): 412–24.
Botting, F., 'The art of smoking in an age of techno-moral consumption', *New Formations*, 38 (1999): 78–97.
Brasher, P. H., 'Clinical and social problems associated with familial intestinal polyposis', *Archives of Surgery*, 69 (1954): 785–96.
Bunting, M., 'Diving into the unknown', *Guardian*, 12 June 2000, p. 17.
Burchell G. *et al.* (eds), *The Foucault Effect: Studies in Governmentality* (Chicago: University of Chicago Press, 1991).
Burian, R. M., J. Gayon and D. Zallen, 'The singular fate of genetics in the history of French biology, 1900–1940', *Journal of the History of Biology*, 21 (1988): 357–402.
Butler, J., 'Revisiting bodies and pleasures', in V. Bell (ed.), *Performativity and Belonging* (London: Routledge, 1999), pp. 11–20.
Cantor, D., 'The definition of radiobiology: The Medical Research Council's support for research into biological effects of radiation in Britain, 1919–1939', Ph.D. dissertation, Lancaster University, 1987.
—— 'The Name and the Word: Neo-hippocratism and language in interwar Britain', in Cantor (ed.) *Reinventing Hippocrates* (Aldershot: Ashgate, 2002), pp. 280–301.
Cavanaugh, W. T., 'A fire strong enough to consume the house': The wars of religion and the rise of the state', *Modern Theology*, 11 (1995): 397–420.
Caygill, H., 'Drafts for a methaphysics of the gene', *Tekhnema*, 3 (1996): 141–52.
—— 'Liturgies of fear: Biotechnology and culture', in B. Adam *et al.* (eds), *The Risk Society and Beyond: Critical Issues for Social Theory* (London: Sage, 2000), pp. 155–64.
—— 'Surviving the inhuman', in S. Brewster *et al.* (eds), *Inhuman Reflections: Thinking the Limits of the Human* (Manchester: Manchester University Press, 2000), pp. 217–29.
Clarke, A. E. and J. H. Fujimura (eds), *The Right Tools for the Job: At Work in Twentieth-Century Life Sciences* (Princeton: Princeton University Press, 1992).

Cooke, G. W. (ed.), *Agricultural Research: A History of the Agricultural Research Council and a Review of Developments in Agricultural Science During the Last Fifty Years* (London: ARC, 1981).
Cooter, R., 'Anticontagionism and history's medical record', in Wright and Treacher, *The Problem of Medical Knowledge*, pp. 87–108.
—— 'The resistible rise of medical ethics', *Social History of Medicine*, 8 (1995): 257–70.
—— 'The ethical body', in Cooter and J. Pickstone (eds), *Medicine in the 20th Century* (Amsterdam: Harwood Academic Press, 2000), pp. 451–68.
Cramer, W. and E. S. Horning, 'Adrenal changes associated with oestrin administration and mammary cancer', *Journal of Pathology*, 44 (1937): 633–42.
Dale, H. E., *Daniel Hall: Pioneer in Scientific Agriculture* (London: Murray, 1956).
Dampier, W. C., 'Agricultural research and the work of the Agricultural Research Council', *Journal of the Farmers' Club* (1938): 55–61.
Dawkins, R., 'The word made flesh', *Guardian*, 27 December 2001, p. 11.
De Certeau, M., *The Mystic Fable: The Sixteenth and Seventeenth Centuries* [1982] (Chicago: University of Chicago Press, 1992).
De Jager, T., 'Pure science and practical interests: The origins of the Agricultural Research Council, 1930–1937', *Minerva*, 31 (1993): 129–50.
De Man, P., *Blindness and Insight: Essays in the Rhetoric of Contemporary Criticism* (London: Routledge, 1983).
Deleuze, G., *Difference and Repetition* [1968] (London: Athlone, 1994).
Deleuze, G. and F. Guattari, *Anti-Oedipus: Capitalism and Schizophrenia* [1973] (London: Athlone Press, 1984).
—— *A Thousand Plateaus: Capitalism and Schizophrenia* [1980] (London: Athlone, 1988).
Derrida, J., 'Cogito and the history of madness' [1964], in Derrida, *Writing and Difference* (London: Routledge, 1978), pp. 31–63.
—— *Of Grammatology* [1967] (Baltimore: Johns Hopkins University Press, 1976).
—— *The Gift of Death* (Chicago: University of Chicago Press, 1995).
—— *Archive Fever: A Freudian Impression* (Chicago: University of Chicago Press, 1996).

Desmond, A. J. and J. Moore, *Darwin: The Life of a Tormented Evolutionist* (New York: Warner Books, 1991).
Dillon, M., 'Another justice', *Political Theory*, 27 (1999): 155–75.
Douglas, K., 'Replaying life', *New Scientist*, 13 February 1999, pp. 29–33.
Drummond, M., 'Report of the research director', *Annual Report of the Scottish Plant Breeding Station* (1922): 13–19.
Duden, B., *The Woman Beneath the Skin: A Doctor's Patients in Eighteenth-Century Germany* (Cambridge, Mass.: Harvard University Press, 1991).
Dukes, C. E., 'Familial intestinal polyposis', *Annals of Eugenics*, 17 (1952): 1–29.
Edgerton, D., 'The "white heat" revisited: The British government and technology in the 1960s', *Twentieth Century British History*, 7 (1996): 53–82.
Engelhardt, D. v., 'Medicina e letteratura', *L'Arco di Giano*, 21 (1999): 157–65.
Engledow, F. L., 'Rowland Harry Biffen', *Obituary Notices of Fellows of the Royal Society*, 7 (1950–51): 9–25.
Faubion, J. (ed.), *Michel Foucault: Essential Works of Foucault 1954–1984* (London: Penguin, 1998–2000), 3 vols.
Fissell, M. E., *Patients, Power and the Poor in Eighteenth Century Bristol* (Cambridge: University of Cambridge Press, 1991).
Foucault, M., *Madness and Civilization: A History of Insanity in the Age of Reason* [1961] (New York: Vintage Press, 1973).
—— *The Order of Things: An Archaeology of the Human Sciences* [1966] (New York: Vintage Press, 1973).
—— 'Politics and the study of discourse [1968]', in Burchell, *The Foucault Effect*, pp. 53–72.
—— 'What is an author?' [1969], in Faubion, *Michel Foucault*, Vol. 2, pp. 205–22.
—— *Discipline and Punish: The Birth of the Prison* [1975] (London: Penguin, 1991).
—— *History of Sexuality* [1976] (London: Penguin, 1990).
—— 'Questions of method' [1977], in Burchell, *The Foucault Effect*, pp. 73–86.
—— 'Truth and power' [1977], in C. Gordon (ed.), *Power/Knowledge: Selected Interviews and Other Writings by Michel Foucault* (New York: Prentice-Hall, 1980), 109–33.

—— 'What is enlightenment?' [n.d.], in Faubion, *Michel Foucault*, Vol. 1, pp. 303–19.

Fox Keller, E., *A Feeling for the Organism: The Life and Times of Barbara McClintock* (New York: Freeman, 1983).

—— *The Century of the Gene* (Cambridge, Mass.: Harvard University Press, 2000).

Franklin, S., 'Gene spawns a lot of questions', *Times Higher Education Supplement*, 15 June 2001, p. 19.

Freedland, J., 'Goodbye to the oracle', *Guardian*, 9 June 1999, p. 19.

Frye, M., *The Politics of Reality: Essays in Feminist Theory* (Freedom: Crossing Press, 1983).

Gaudillière, J.-P., 'Circulating mice and viruses: the Jackson Memorial Laboratory, the National Cancer Institute, and the genetics of breast cancer, 1930–1965', in M. Fortun and E. Mendelsohn (eds), *The Practices of Human Genetics* (Dordrecht: Kluwer, 1999), pp. 89–124.

—— and I. Löwy, 'Disciplining cancer: Mice and the practice of genetic purity', in Gaudillière and Löwy (eds), *The Invisible Industrialist: Manufactures and the Production of Scientific Knowledge* (London: Macmillan, 1998), pp. 209–49.

Geison, G., *The Private Life of Louis Pasteur* (Princeton: Princeton University Press, 1994).

Ginzburg, C., 'Clues: Roots of an evidential paradigm', in Ginzburg, *Myths, Emblems, Clues* (London: Hutchinson Radius, 1986), pp. 96–125.

Gould, S. J., *The Panda's Thumb: More Reflections in Natural History* (New York: Norton, 1982).

Gumbrecht, H. U., 'Perception versus experience', in Lenoir, *Inscribing Science*, pp. 341–64.

Haldane, J. B. S., 'The prospects of eugenics', in M. L. Johnson, M. Abercrombie and G. E. Fogg (eds), *New Biology* (London: Penguin, 1956), Vol. 22, pp. 7–23.

Hall, A. D., *A Pilgrimage of British Farming, 1910–1912* (London: Murray, 1913).

Haraway, D., *Modest Witness @ Second Millennium: FemaleMan© Meets OncoMouse™* (London: Routledge, 1997).

Hardt, M. and A. Negri, *Empire* (Cambridge, Mass.: Harvard University Press, 2000).

Harwood, J., *Styles of Scientific Thought: The German Genetics Commu-*

nity 1900–1933 (Chicago: University of Chicago Press, 1993).
Heidegger, M., 'The question concerning technology' [1954], in *The Question Concerning Technology* (New York: Garland, 1977), pp. 3–49.
Hopkins, Sir Frederick Gowland, 'The clinician and the laboratory worker' [1931], in J. Needham and E. Baldwin (eds), *Hopkins and Biochemistry* (Cambridge: Heffer, 1949), pp. 206–10.
Horrocks, S. M., 'Nutrition science and the food and pharmaceutical industries in interwar Britain', in D. Smith (ed.), *The History of Nutrition: Institutional, Professional, Scientific and Policy Issues* (London: Routledge, 1997), pp. 53–74.
Hughes, T. P., *Networks of Power: Electrification in Western Society, 1880–1930* (Baltimore: Johns Hopkins University Press, 1983).
Hunter, H., 'Development in plant-breeding', in Sir Daniel Hall (ed.), *Agriculture in the Twentieth Century* (Oxford: Clarendon Press, 1939), pp. 223–60.
Jameson, F., *Postmodernism, or the Cultural Logic of Late Capitalism* (London: Verso, 1991).
Jinks, J. (ed.), *Fifty Years of Genetics: Proceedings of a Symposium held at the 160th Meeting of the Genetical Society of Great Britain on the 50th Anniversary of its Foundation* (Edinburgh: Oliver and Boyd, 1970).
Kamminga, H. and M. W. Wetherall, 'The making of a biochemist, I: Frederick Gowland Hopkins' construction of dynamic biochemistry', *Medical History*, 40 (1996): 269–92.
—— and —— 'The making of a biochemist, II: The construction of Frederick Gowland Hopkins' reputation', *Medical History*, 40 (1996): 415–36.
Kargon, R., *Science in Victorian Manchester: Enterprise and Expertise* (Manchester: Manchester University Press, 1977).
Kay, L. E., *Who Wrote the Book of Life? A History of the Genetic Code* (Stanford: Stanford University Press, 2000).
Kirkwood, T., *Times of our Lives: The Sciences of Human Ageing* (Oxford: Oxford University Press, 2000).
Kuhn, T. S., *The Structure of Scientific Revolutions* (Chicago: University of Chicago Press, 1962).
Lander, E. S. *et al.*, 'Initial sequencing and analysis of the human genome', *Nature*, 409 (2001): 860–921.
Latour, B., *The Pasteurization of France* (Cambridge, Mass.: Harvard Uni-

versity Press, 1988).
—— 'The force and the reason of experiment', in H. E. Le Grand (ed.), *Experimental Inquiries: Historical, Philosophical and Social Studies of Experimentation in Science* (Dordrecht: Kluwer, 1990), pp. 49–80.
—— *Pandora's Hope: Essays on the Reality of Science Studies* (Cambridge, Mass.: Harvard University Press, 1999).
—— *Politiques de la Nature: Comment Faire Entrer les Sciences en Démocratie* (Paris: La Découverte, 1999).
—— and S. Woolgar, *Laboratory Life: The Construction of Scientific Facts [2nd edn]* (Princeton: Princeton University Press, 1986).
Lawrence, C., *Medicine in the Making of Modern Britain, 1700–1920* (London: Routledge, 1994).
—— 'Still incommunicable: Clinical holists and medical knowledge in interwar Britain', in Lawrence and G. Weisz (eds), *Greater Than the Parts: Holism in Biomedicine, 1920–1950* (Oxford: Oxford University Press, 1998), pp. 94–112.
Le Goff, J., *The Birth of Purgatory* (London: Scolar Press, 1984).
—— *Your Money or Your Life: Economy and Religion in the Middle Ages* (New York: Zone, 1988).
Lecourt, D., *Contre la Peur: de la Science à l'Éthique, une Aventure Infinie* (Paris: Hachette, 1990).
Lenoir, T. (ed.), *Inscribing Science: Scientific Texts and the Materiality of Communication* (Stanford: Stanford University Press, 1998).
—— 'Inscription practices and materialities of communication', in Lenoir, *Inscribing Science*, pp. 1–19.
Levinas, E., 'Ethics as first philosophy [1984]', in S. Hand (ed.), *The Levinas Reader* (Oxford: Blackwell, 1989), pp. 76–87.
Lockhart Mummery, J. P., 'The diagnosis of tumours in the upper rectum and sigmoid flexure by means of the electric sigmoidoscope', *Lancet*, (1904), Vol. 1: 1781–2.
—— *Diseases of the Rectum and Colon and their Surgical Treatment* (London: Ballière, 1923).
—— 'The Royal Society discussion on experimental production of malignant tumours', *Lancet* (1933), Vol. 2: 323.
—— 'Prevention of cancer', *Lancet* (1934), Vol. 1: 155.
—— *The Origin of Cancer* (London: Churchill, 1934).
—— 'Medical science and social progress', *British Medical Journal* (1935),

Vol. 2: 1022.
—— *After Us, or the World as it Might Be* (London: Stanley Paul, 1936).
—— 'The surgeon as a biologist', *Surgery, Gynaecology, and Obstetrics*, 66 (1938): 257–63.
—— 'Summary', *Annual Report of the British Empire Cancer Campaign*, 17 (1940): 8–21.
—— *Nothing New under the Sun* (London: Andrew Melrose, 1947).
Lynch, M., 'Archives in formation: Privileged spaces, popular archives and paper trails', *History of the Human Sciences*, 12 (1999): 65–87.
Mackenzie, D. and S. B. Barnes, 'Scientific judgment: The biometry-mendelism controversy', in Barnes and S. Shapin (eds), *Natural Order: Historical Studies in Scientific Culture* (London: Sage, 1979), pp. 191–210.
Marx, K. and F. Engels, 'Manifesto of the Communist Party [1848]', in R. C. Tucker (ed.), *The Marx-Engels Reader* (New York: Norton, 1972), pp. 335–62.
Meek, J., 'Gene test plea to cut cancer of bowel risk', *Guardian*, 27 June 2001, p. 5.
Monod, J., *The Statue Within: An Autobiography* (New York: Basic Books, 1988).
Moreira, T., 'Translation, difference and ontological fluidity: Cerebral angiography and neurosurgical practice, 1926–45', *Social Studies of Science*, 30 (2000): 421–46.
Moynihan, Lord, 'The science of medicine', *Lancet* (1930), Vol. 2: 779–85.
Mulgan, R. G., 'Aristotle's doctrine that man is a political animal', *Hermes* 102 (1974): 438–45.
Nietzsche, F., 'On the uses and disadvantages of history for life' [1874], in *Untimely Meditations* (Cambridge: University of Cambridge Press, 1983), pp. 59–123.
Northover, J. M. A., 'Imperial Cancer Research Fund colorectal cancer unit', *St Mark's Hospital for Diseases of the Rectum and Colon, Annual Report* (1984): 53–4.
Olby, R. C., *The Origins of Mendelism [2nd edn]* (Chicago: University of Chicago Press, 1985).
—— 'Scientists and bureaucrats in the establishment of the John Innes Horticultural Institution under William Bateson', *Annals of Science*, 46 (1989): 497–510.
—— 'Social imperialism and state support for agricultural research in

Edwardian Britain', *Annals of Science*, 48 (1991): 509–26.
Pal, B. P. and D. K. Mukherji, 'Sir Rowland Biffen: A pioneer in genetic research', *Indian Journal of Genetics and Plant Breeding*, 9 (1950): 84–5.
Palladino, P., 'Sterochemistry and the nature of life: Mechanist, vitalist and evolutionary perspectives', *Isis*, 81 (1990): 44–67.
—— 'The political economy of applied science: Plant breeding research in Great Britain, 1910–1940', *Minerva*, 28 (1990): 446–68.
—— 'Between craft and science: Plant breeding, mendelian genetics, and British universities, 1900–1920', *Technology and Culture*, 34 (1993): 300–23.
—— 'Wizards and devotees: On the mendelian theory of inheritance and the professionalization of agricultural science in Great Britain and the United States, 1880–1930', *History of Science*, 32 (1994): 409–44.
—— 'Science, technology, and the economy: Plant breeding in Great Britain, 1920–1970', *Economic History Review*, 48 (1995): 116–36.
—— 'The empire, colonies and lesser developed countries as mirror: Critical reflections on science for economic development in the colonial and post-colonial periphery, 1930–1970', in Y. Chatelin and C. Bonneuil (eds), *Nature et Environnement* (Paris: ORSTOM, 1996), pp. 243–53.
—— 'On writing the histor(ies) of modern medicine', *Rethinking History*, 3 (1999): 271–88.
—— 'Icarus' flight: On the dialogue between the historian and the historical actor', *Rethinking History*, 4 (2000): 21–36.
—— 'Speculations on cancer-free babies: Surgery and genetics at St Mark's Hospital, 1924–1995', in J.-P. Gaudillière and I. Löwy (eds), *Heredity and Infection: The History of Disease Transmission* (London: Routledge, 2001), pp. 285–310.
—— 'Between knowledge and practice: On medical professionals, patients and the making of the genetics of cancer', *Social Studies of Science*, 32 (2002), 137–65.
Passey, R. D., 'Cancer', *Journal of the Royal Sanitary Institute*, 47 (1927): 653.
Percival, J., *The Wheat Plant* (London: Duckworth, 1921).
Pirandello, L., 'Six characters in search of an author [1921]', in R. Rietty *et al.* (trans.), *Pirandello: Three Plays* (London: Methuen, 1985), pp. 69–134.
Ponge, F., 'My creative method' [1947], in *Le Grand Recueil: Méthodes*

(Paris: Gallimard, 1961), pp. 12–13.
Porter, R. S., 'Introduction', in Porter (ed.), *Patients and Practitioners: Lay Perceptions of Medicine in Pre-Industrial Society* (Cambridge: University of Cambridge Press, 1985), pp. 1–22.
Provine, W. B., *The Origins of Theoretical Population Genetics* (Chicago: University of Chicago Press, 1971).
Rabeharisoa V. and M. Collan, *Le Pouvoir des Malades: l'Association Française contre les Myopathies et la Recherche* (Paris: École des Mines de Paris, 1999).
Rabinow, P., 'Artificiality and enlightenment: From sociobiology to biosociality', in Rabinow, *Essays on the Anthropology of Reason* (Princeton: Princeton University Press, 1996), pp. 91–111.
—— *French DNA: Trouble in Purgatory* (Chicago: University of Chicago Press, 1999).
—— 'Epochs, presents, events', in M. Lock *et al.* (eds), *Living and Working with the New Medical Technologies* (Cambridge: Cambridge University Press, 2000), pp. 31–46.
Rancière, J., *Dis-agreement: Politics and Philosophy* (Minneapolis: University of Minnesota Press, 1998).
Rella, F., 'Critica e storia', in Rella (ed.), *Critica e Storia: Materiali su Benjamin* (Venice: Cluva, 1980), pp. 9–29.
Rheinberger, H.-J., 'Experimental systems, graphematic spaces', in Lenoir, *Inscribing Science*, pp. 283–303.
Ritvo, H., *The Animal Estate: The English and Other Creatures in the Victorian Age* (Cambridge, Mass.: Harvard University Press, 1987).
—— *The Platypus and the Mermaid, and Other Figments of the Classifying Imagination* (Cambridge, Mass.: Harvard University Press, 1997).
Rogers, L., 'Doctors to create cancer-free babies', *The Sunday Times*, 5 November 1995, p. 24.
Rose, D., 'I want to scream with rage at this', *Observer*, 15 April 2001, pp. 14–15.
Rose, G., *Mourning Becomes the Law: Philosophy and Representation* (Cambridge: University of Cambridge Press, 1996).
Salaman, R. N., *The History and Social Influence of the Potato* (Cambridge: Cambridge University Press, 1949).
Samuel, R., *Theatres of Memory: Past and Present in Contemporary Culture* (London: Verso, 1994)

—— 'The people with stars in their eyes', in A. Light (ed.), *Island Stories: Unravelling Britain* (London: Verso, 1998), pp. 224–9.
Schofield, M., *Saving the City: Philosopher-Kings and Other Classical Paradigms* (London: Routledge, 1999).
Scholem, G., *Walter Benjamin: Die Geschichte einer Freundschaft* (Frankfurt: Suhrkamp, 1975).
Scott, J. W. (ed.), *Feminism and History* (New York: Oxford University Press, 1996).
Segerstrale, U., *Defenders of the Truth: The Sociobiology Debate* (Oxford: Oxford University Press, 2001).
Shakespeare, T., 'Back to the future? New genetics and disabled people', *Critical Social Policy*, 46 (1995): 22–35.
Shiva, V., *The Violence of the Green Revolution: Ecological Degradation and Political Conflict* (London: Zed Books, 1991).
Smith, D., 'The use of "team work" in the practical management of research in the inter-war period: John Boyd Orr at the Rowett Research Institute', *Minerva*, 37 (1999): 259–94.
Smocovitis, V. B., 'Living with your biographical subject: Special problems of distance, privacy and trust in the biography of G. Ledyard Stebbins Jr', *Journal of the History of Biology*, 32 (1999): 421–38.
Söderquist, T., 'Existential projects and existential choice in science: Science biography as an edifying genre', in R. Yeo and M. Shortland (eds), *Telling Lives: Studies of Scientific Biography* (Cambridge: University of Cambridge Press, 1995), pp. 45–84.
—— *What Struggle to Escape: The Life and Science of Niels Jerne* (New Haven: Yale University Press, forthcoming, 2003).
Squier, S. M., *Babies in Bottles: Twentieth-Century Visions of Reproductive Technologies* (New Brunswick: Rutgers University Press, 1994).
Stapledon, Sir George, *The Way of the Land* (London: Faber and Faber, 1943).
Steedman, C., *Past Tenses: Writing, Autobiography and History* (London: Rivers Oram, 1992).
—— *Dust* (Manchester: Manchester University Press, 2001).
Sturdy, S., 'The germs of a new Enlightenment', *Studies in the History and Philosophy of Science*, 22 (1991): 163–73.
______ and R. Cooter, 'Science, scientific management, and the transformation of medicine in Britain, 1870–1950', *History of Science*, 31 (1998): 421–66.

Tait, N. and M. Urry, 'Monsanto pays £320m for UK crop breeding business', *Financial Times*, 16 July 1998, p. 33.

Thirtle, C. and V. W. Ruttan, *The Role of Demand and Supply in the Generation and Diffusion of Technical Change* (Chur: Harwood Academic Press, 1987).

——, P. Palladino and J. Piesse, 'On the organisation of agricultural research in the United Kingdom: A quantitative description and appraisal of recent reforms', *Research Policy*, 26 (1997): 557–76.

Traweek, S., 'Border crossings: Narrative strategies in science studies and among physicists in Tsukuba Science City, Japan', in A. Pickering (ed.), *Science as Practice and Culture* (Chicago: University of Chicago Press, 1992), pp. 429–65.

Turney, J., *Frankenstein's Footsteps: Science, Genetics and Popular Culture* (New Haven: Yale University Press, 1998).

Veale, A. M. O., 'Genetics, carcinogenesis, and family studies', *British Surgical Practice and Surgical Progress* (1961): 169–85.

Velody, I., 'The archive and the human sciences: Notes towards a theory of the archive', *History of the Human Sciences*, 11 (1998): 1–16.

Vines, G., 'Star of the big screen', *Times Higher Education Supplement*, 21 May 1996, p. 14.

Virilio, P., *Open Sky* (London: Verso, 1997).

—— *The Information Bomb* (London: Verso, 2000).

Warner, J. H., 'The history of science and the sciences of medicine', *Osiris*, 10 (1995): 164–93.

Webster, A. J., 'Privatisation of public sector research: The case of a plant-breeding institute', *Science and Public Policy*, 16 (1989): 224–32.

Wiesing, U., 'Die Einsamkeit des Artzes und der 'lebendige Drang nach Geschichte'. Zum historischen Selbverständnis der Medizin bei Richard Koch', *Gesnerus*, 54 (1997): 219–41.

Wilkins, V. E., *Research and the Land: An Account of Recent Progress in Agricultural and Horticultural Science in the United Kingdom* (London: HMSO, 1926).

Williams, R., *Marxism and Literature* (Oxford: Oxford University Press, 1977).

Wood, T. B., 'The School of Agriculture of the University of Cambridge', *Journal of the Ministry of Agriculture*, 29 (1922): 223–30.

Woolgar, S., 'Interests and explanation in the social study of science',

Social Studies of Science, 11 (1981): 365–94.
Wright, P. and A. Treacher (eds), *The Problem of Medical Knowledge: Examining the Social Construction of Medicine* (Edinburgh: Edinburgh University Press, 1982)
Wyschogrod, E., *An Ethics of Remembering: History, Heterology, and Nameless Others* (Chicago: University of Chicago Press, 1998).
Yeats, W. B., 'The second coming' [1921], in *The Collected Poems of W. B. Yeats* (London, 1950), pp. 210–11.
Young, R. M., 'Desmond and Moore's Darwin: A critique', *Science as Culture*, 4 (1994): 393–424.
Žižek, S., *The Ticklish Subject: The Absent Centre of Political Ontology* (London: Verso, 1999).
—— *The Fragile Absolute: Or, Why the Christian Legacy is Worth Fighting For?* (London: Verso, 2000).

Index